2012 System, Software, SoC and Silicon Debug Conference

(S4D 2012)

Vienna, Austria
19 – 20 September 2012

IEEE Catalog Number: CFP1237T-PRT
ISBN: 978-1-4673-2454-0

Copyright © 2012, Electronic Chips & Systems Design Initiative
All Rights Reserved

***This publication is a representation of what appears in the IEEE Digital Libraries. Some format issues inherent in the e-media version may also appear in this print version.

IEEE Catalog Number: CFP1237T-PRT
ISBN 13:   978-1-4673-2454-0
ISSN: 2114-3684

Additional Copies of This Publication Are Available From:

Curran Associates, Inc
57 Morehouse Lane
Red Hook, NY 12571 USA
Phone: (845) 758-0400
Fax: (845) 758-2633
E-mail: curran@proceedings.com
Web: www.proceedings.com

2012 System, Software, SoC and Silicon Debug Conference (S4D 2012)

Vienna, Austria
19-20 September 2012

IEEE Catalog Number: CFP1237T-POD
ISBN: 978-1-46732-454-0

Table of Contents

 s4d

Welcome 4

Chairs 4

Keynote Speakers 5

Program Committee 5

Local Organizer 6

Sponsors 6

Demonstrations 6

Conference Papers 7

Session 1: Software Instrumentation 7

Minimising the Impact of Software Instrumentation Using On-chip Debug and a Secondary CPU Core... 8
  Padraig Fogarty (University of Limerick)

Software Instrumentation of Safety Critical Embedded Systems - a Problem Statement........................ 13
  Mathias Ekman and Henrik Thane (Mälardalen University)

Session 2: Profiling, Runtime Verification, Reverse Debug & Speedpath Diagnosis 17

Profiling Bare-Metal Cores in AMP Systems 18
  Adriaan Schmidt (Fraunhofer ESK)

A Runtime Verification Unit for Microcontrollers 22
  Thomas Reinbacher, Andreas Steininger (Vienna University of Technology),
  and Martin Horauer (University of Applied Sciences Technikum Wien)

A Review of Reverse Debugging 28
  Jakob Engblom (Wind River)

Application of Timing Variation Modeling to Speedpath Diagnosis.. 34
  Mehdi Dehbashi and Görschwin Fey (University of Bremen)

Session 3: Verification and Virtual Prototyping.. 38

Co-Debug and Co-Verification Environment for Power Management System 39
  Markus Winterholer (Cadence)

Scalable and Retargetable Debugger Architecture for Heterogeneous MPSoCs 45
  Luis Gabriel Murillo, Julian Harnath, Rainer Leupers and Gerd Ascheid (RWTH Aachen)

Co-Simulation Framework for Variation Analysis of Radio Frequency Transceivers............................ 51
  Sumit Adhikari, Florian Schupfer and Christoph Grimm (Vienna University of Technology)

Session 4: Tracing.. 57

Compact Function Trace (CFT).. 58
  Albrecht Mayer and Reinhard Deml (Infineon)

CULT: A Unified Framework for Tracing and Logging C-based Designs 60
  Wei Hong, Alexander Viehl (Forschungszentrum Informatik Karlsruhe), Nico Bannow,
  Christian Kerstan, Hendrik Post (Robert Bosch GmbH), Oliver Bringmann and Wolfgang Rosenstiel
  (University of Tuebingen)

ARMv8 Debug and Trace Architectures 66
  Michael Williams, ARM Limited

Welcome to S4D 2012

Embedded systems and software complexity is rapidly increasing; from one processor to ten or more, from thousands of lines of code to millions. Modern system-on-chips integrate dozens of hardware accelerators and I/O devices. Modern products integrate hard IP and software from many more sources than in the recent past. Embedded systems debug capabilities must scale with this exponential growth in system complexity.

This fourth edition of S4D Conference has evolved from several industry workshops organized by ECSI into the areas of debug and provides a forum for work and standardization efforts related to debug of electronic systems. The conference addresses a vast set of requirements from system and SoC companies with regard to debug methods and tools. It includes efforts from IEEE and other standards working groups, including the Nexus 5001 Forum, IJTAG (IEEE P1687), IEEE 1149.7, MIPI, industry working groups including OCP-IP, the Multicore Association (MCA), the Accellera - SPIRIT Consortium and Eclipse and industry R&D projects contributing to debug and related tools and methods development. The S4D Conference event also allows presentation and discussion of existing and new commercial debug tools and products related to electronic (silicon and software) methods, devices, and systems. In summary, the S4D Conference provides a forum for discussing research, scientific and commercial development in the areas of system, software, SoC and silicon debug.

Conference Chairs

General Chair: Peter Rössler, University of Applied Sciences, Technikum Wien

Peter Rössler received his Diploma and Ph.D. from the Vienna University of Technology. Currently he works as a FH-Professor at the University of Applied Sciences Technikum Wien in Vienna/Austria where he is managing R&D projects in the area of embedded systems. His main research interests cover topics like FPGA, ASIC and ESL design as well as control networks (fieldbus systems). He is co-founder of a company related to building automation systems. Peter Rössler is author of 50+ publications. He is member of conference program committees like ASME/IEEE MESA, Austrochip or ME and is engaged in boards of national organizations like the Austrian Electrotechnical Association (ÖVE) or the Austrian Society of Microelectronic Systems. Since 2012 he is Chair of the IEEE Austrian Section.

Co-Chair: Adam Morawiec, ECSI

Adam Morawiec received his MSc degree in electronic system design in 1993 from the Silesian Technical University in Gliwice, Poland and his DEA (Diplome d'Etudes Approffindies) in 1996 and PhD in 2000 in Microelectronics at the TIMA Laboratory / Université Joseph Fourrier, Grenoble, France in the domain of verification and simulation performance methods. He works for in the R&D project development and management in the domain of system design methods and standards, in setting up industry and research consortia, in organization of advanced training and workshop in system design area. He also acted as an expert of the European Commission in the R&D project proposal evaluation and IST/ICT Work programmes definition. Since 2005 he is the director of ECSI. He is an author of several scientific publications in the area of formal verification, formal models, simulation performance and system design. He is also an editor of two technical books published by Springer Publishers: "Platform Based Design at the Electronic System Level" and "High-Level Synthesis".

Keynote Speakers

Albrecht Mayer, Infineon

Albrecht Mayer is Senior Principal Emulation Systems and Tooling at Infineon. Within the automotive microcontroller business unit he is responsible for on-chip debug architectures and C-modeling methodology. He defined with his team the TriCore™ debug architecture, which is the benchmark in the automotive industry. He has published many papers and holds more than 20 patents. Dr. Mayer received Diploma and PhD degrees in electrical engineering from the Technical University of Munich.

Ingo Sander, KTH Royal Institute of Technology

Ingo Sander received the MSc degree in Electrical Engineering from the Technical University of Braunschweig, Germany, in 1990 and the PhD degree from KTH - Royal Institute of Technology, Sweden, in 2003. Between 1991 and 1993 he has worked as system design engineer at Ericsson, Sweden. In 1993 he joined KTH, where he since 2005 holds a position as associate professor in Electronic System Design. His main research interests are located in the area of design methodologies for embedded systems. His current research is aiming towards predictable performance of real-time applications on multi-processor platforms by integration of formal models into the design flow.

Program Committee

Tapani Ahonen	Tampere U. of Technology	Klaus McDonald-Maier	U. of Essex/UltraSoC
Jens Braunes	PLS Development Tools	Adam Morawiec	ECSI
Pat Brouillette	Roku	Brenden Mullane	U. of Limerick
Philippe Cuenot	Continental Automotive	Chris Ng	IBM
Serge De Paoli	STMicroelectronics	Frédéric Petrot	TIMA Laboratory
Jakob Engblom	Wind River	Paolo Prinetto	Politecnico di Torino
Philipp Graf	FZI	David Riemens	NXP
Jan Haase	TU Vienna	Neal Stollon	HDL Dynamics
Ziyad Hanna	Jasper Design Automation	Erich Styger	Freescale
Andreas Hoffmann	Synopsys	Bart Vermeulens	NXP
Rolf Kühnis	Intel	Alexander Weiss	Accemic
Rainer Leupers	RWTH AACHEN University	Michael Williams	ARM
Andrea Martin	Lauterbach	Markus Winterholer	Cadence
Albrecht Mayer	Infineon	Hans-Joachim Wunderlich	U. of Stuttgart

Local Organizer

TECHNISCHE
UNIVERSITÄT
WIEN
Vienna University of Technology

Demonstrations

Sponsor

System Engineering Tool and UML-Debugger for Enterprise Architect from Sparx Systems
LieberLieber AMUSE is a simulation tool for UML or SysML and a model debugger. The product is the result of our rich experience gained in various system engineering projects over the last years. We strongly believe that the model-based software and system engineering will determine the future of the industry sector and we are proud to support the trend with our products.

 Vienna, Austria – September 19-20, 2012

Session 1: Software Instrumentation

Minimising the Impact of Software Instrumentation Using On-chip Debug and a Secondary CPU Core
  Padraig Fogarty (University of Limerick)

Software Instrumentation of Safety Critical Embedded Systems - a Problem Statement
  Mathias Ekman and Henrik Thane (Mälardalen University)

www.ecsi.org/s4d

MINIMISING THE IMPACT OF SOFTWARE INSTRUMENTATION USING ON-CHIP DEBUG AND A SECONDARY CPU CORE

Padraig Fogarty
Department of Electronic and Computer Engineering,
University of Limerick, Limerick, Ireland
padraig.fogarty@ul.ie

ABSTRACT

As SoC and general purpose controllers become more complex and more tightly integrated, gaining access to the runtime data required for software verification and debug has become more challenging. This means software engineers are now more reliant than ever upon software instrumentation and the on-chip mechanisms provided to extract this vital information. In this paper the author proposes using on-chip debug hardware coupled with a secondary CPU core to extract instrumentation data. Using a secondary core alleviates the need to modify the source code on the primary CPU and therefore avoids many of the difficulties with modifying code and minimises the impact of the software instrumentation.

Index Terms— Embedded software, Real time systems, Software debugging, Software tools, Software testing.

1. INTRODUCTION

In the embedded system design it is necessary to extract data regarding software execution from the target; either for verification or debug purposes. As chip architectures become more complex, so too does the challenge of accessing this debug and verification data. In the past it was possible to monitor software execution using emulation systems or logic analysers attached to external buses and I/O signals. However, the introduction of multicore SoC architectures has created a scenario whereby the data and signals to be monitored are no longer externally visible.

Most SoC designs include some form of on-chip test or debug interface, and where possible, it is good practice to reuse the same I/O interface and on-chip circuitry for both purposes. Indeed, there are many common objectives for test and debug, not least the desire to simplify and speed-up the extraction of data at runtime; and many standard and proprietary solutions aim to address these needs [1], [2], [3], [4], [5]. Industry standards focussed at application specific needs have had varying degrees of success; IEEE-ISTO 5001 – 2003 (Nexus) [6] which is aimed at automotive applications has not been widely adopted, whereas, the debug specification from the Mobile Industry Processor Interface (MIPI) Alliance [7] appears to be well established.

Ultimately, economic constraints dictate the finite resources that can be dedicated to test and debug functions, and in most cases it is no longer possible to monitor and extract all noteworthy on-chip activity using the limited hardware features provided. Mayer et al. [4] address this challenge in a two-fold manner. Firstly, by providing complex on-chip hardware triggers to better isolate the data which is of interest; this also reduces the amount of data which must be sent off-chip. Secondly, the debug circuitry is designed to sit outside of the SoC layout area, either on the periphery or on a separate die; which eases the removal of the debug circuitry if it is not required for volume production.

Given the cost and complexity of adding dedicated debug capabilities, using software instrumentation offers an attractive alternative. Unfortunately, on-chip test and debug interfaces are not always available or suitable for software instrumentation purposes. It is therefore common practice to output software instrumentation data using a general purpose serial or similar communications interface. One advantage of using a standard communications interface and protocol is the relative ease with which data can be outputted in a human readable format. This not only simplifies the process of checking the output data for engineers, but is particularly important if a transcript of the output must be retained for quality systems audit purposes. A thorough external auditor will follow the entire trail from quality systems procedures through to, and including, output data records demonstrating compliance with these procedures. If an additional tool is required to interpret the output data, then the verification and validation of this tool will also be subject to rigorous audit; consequently minimising the number of tools required reduces potential sources of error and also streamlines the audit process.

Although convenient and inexpensive, general purpose serial communications interfaces typically have limited bandwidth and often have high latency. An alternative is to use a hybrid of hardware and software designed to ease software instrumentation. The ARM 'CoreSight' suite of modules includes instrumentation blocks, which are intended to speed up and simplify software instrumentation [5]. Brouillette described the 'SVEN and OMAR' platform [8], which enables the capture of software and signal events at high-speed to a memory buffer from which it can later be extracted and analysed. And the MIPI test and debug framework includes a system trace module which can collect trace data from software or hardware sources [7]. Although there are many proprietary solutions for software instrumentation, there is no standard method for formatting, and hence recording, instrumentation or trace data [9]. And while hybrid hardware and software solutions may speed up data capture they still require adding instrumentation to the source code.

This process of adding software instrumentation is not without problems. One difficulty with adding software instrumentation is the overhead it can impose upon the application. If a robust communications protocol is used, the software stack required may be significant, and even a simple *printf* function may require a considerable amount of additional library code. In addition to the extra resources required, this overhead can also impact performance [10], [11]. For cost-sensitive applications or safety-critical embedded control applications this additional overhead may not be acceptable. A further difficulty is that all instrumentation code must be debugged and verified, adding greatly to the development effort. Plus, making any modification to software has the potential to exhibit the classic 'Heisenbug' phenomenon [12]; whereby the instrumented code may mask the bug being observed, or may itself introduce a bug.

The potential to inadvertently introduce an error or bug is possibly the least attractive aspect of software instrumentation. In some limited situations, where resources are not a concern, it may be acceptable for software engineers to simply add instrumentation code as required. However, in well regulated industries, and in particular for safety-critical applications, any change to software whether for application or instrumentation purposes, requires careful management. In these industries, as much emphasis is placed upon the software quality management systems, as upon the development process itself. Where the C language is used for embedded systems development, it is common to find *#include* and *#ifdef* pre-processor directives spread throughout the source code. These can be convenient to conditionally include or exclude instrumentation code fragments at compile time. Including this information in the source code may seem convenient for code reviews and quality system recording purposes. However, in many cases the same directives are also used to conditionally include application code; which can make differentiation between the two difficult. In-house or industry standard guidelines often try to limit usage, or dictate best-practice usage, of such directives [13], [14]. Unfortunately, it is all too easy for a software engineer to simply write a small piece of 'temporary' instrumentation code, and accidently leave this in-situ for the production release.

The possibility of having unnecessary or unknown code included in an embedded system is clearly undesirable from a quality perspective; but it may also pose a serious threat to the reliability, safety, or security, of the system. For verification or test purposes, instrumentation code often includes mechanisms to alter the runtime behaviour of the system. If these mechanisms are not entirely removed or disabled, then the possibility exists that the system behaviour may be accidently or maliciously altered; at best, the consequences may be benign; at worst, they could result in serious injury or loss of life. Like all other source code, software instrumentation requires careful management. At a minimum, thorough code reviews, reliable compilation tools, and robust software configuration management tools are required.

Under ideal circumstances, altering software would not impact upon system resources nor pose any risks. Using existing approaches this is clearly unrealistic for either application or instrumentation code. What is required is a solution leveraging the advantages of instrumentation whilst minimising the disadvantages. The key attributes of such a solution include:

- Low cost, or low incremental cost
- Minimally invasive or non-invasive
- Ability to isolate instrumentation code
- Ability to output data in human readable format
- Ease of use, for software engineers

In the following sections the author outlines the concept of using on-chip debug hardware coupled with a secondary CPU core to extract instrumentation data; using this approach the author believes that many of the problems posed by traditional approaches can be avoided and the key attributes retained. Section 2 provides an outline of the general concept; the demonstration platform used; and examines the results obtained when using this platform to instrument a simple embedded application. In section 3 the author discusses the suitability of this platform when considering the key attributes outline above. And section 4 provides the authors conclusions.

2. CONCEPT

Multiple CPU architectures bring new verification and debug challenges; but these multi-core architectures may also enable a new approach to software instrumentation. On a single CPU device, any instrumentation added to the source code will be invasive to some degree. Whether the overhead required to accommodate and execute this additional instrumentation code is acceptable is very much application dependant. With multiple CPUs the option exists to place all instrumentation code on one CPU, while the application code resides upon the other(s). The challenge then becomes one of executing the appropriate instrumentation code at the appropriate time.

With traditional software instrumentation the code is triggered when execution reaches the point of interest, i.e. the section of code where the instrumentation has been added. With the author's proposed approach, as illustrated in Fig. 1, the instrumented code is activated when the debug hardware detects that execution has reached the same point; the difference being that execution of the instrumented code proceeds on the secondary core.

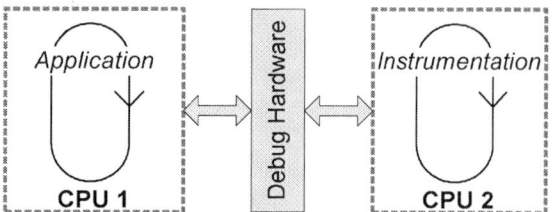

Fig. 1 Instrumentation platform concept

This platform requires that the multicore device has sufficient on chip debug hardware to monitor execution of the application code. But most devices already include some breakpoint or watchpoint capabilities, even if at only a minimal level. However, most on-chip debug hardware is designed to halt execution upon reaching a breakpoint, or capture data for extraction by external tools; therefore what is required, is that the debug hardware be able to trigger execution of the instrumentation code on the secondary on-chip CPU instead. This requirement should not be difficult to achieve and in fact some debug hardware designs already support the ability to trigger on-chip execution.

2.1. Demonstration Platform

A demonstration platform was created to explore the feasibility of the concept proposed. This platform uses the Freescale MC9S12XE microcontroller; which, in addition to a 16-bit CPU12X and peripheral modules also includes an on-chip debug module (S12XDBG) and XGATE coprocessor [15]. These key blocks of the platform are illustrated in Fig. 2 below.

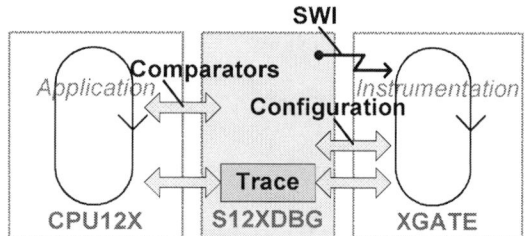

Fig. 2 Demonstration platform using MC9S12XE

The S12XDBG module includes: four comparators, which can monitor the CPU12X or XGATE buses; a state sequencer, which can be used to generate a breakpoint after a certain sequence of comparator triggers; a trace buffer, in which program or memory access information can be captured; and associated control logic. Although tracing of the CPU12X buses is not possible while in background debug mode (BDM), the S12XDBG module can trigger BDM upon completion, which is generally the case. As illustrated in Fig. 2, the trace buffer can capture data from both CPUs, but perhaps more importantly it can also be read by both.

The typical method for configuring the S12XDBG module is to use the BDM interface or to embed debug/monitor software routines on the CPU12X. Both of these approaches are intrusive to any application executing on the CPU12X; entering BDM mode requires halting the CPU12X to configure the debug module; and execution of software functions obviously impacts upon any application executing on the same CPU core. The approach used by the author was to configure the debug module using software on the XGATE coprocessor; thereby not impacting upon application software execution on the CPU12X.

The debug module would typically be configured to generate a breakpoint to BDM, or to trigger a software interrupt (SWI) routine on the CPU12X, which are also intrusive upon application execution. A non-intrusive solution would be to directly trigger a SWI on the XGATE; unfortunately the architecture of the MC9S12XE does not currently support this. Instead, it was necessary to write a two line SWI routine on the CPU12X which in turn triggers the SWI on the XGATE. In this way it was possible to activate the instrumentation code on the XGATE when the relevant point in the application code is reached.

2.2. Results

To evaluate the platform described above the author created a simple real-time application, which generates an analogue waveform using a PWM output from the microcontroller. As with any real-time application the timing of execution is a critical parameter which must be verified to ensure correct operation. Checking the timing of a function or entire application using debug hardware and breakpoints to halt execution may be possible in limited situations; but in general, instrumentation software is added to the application to either use on-chip timing hardware, or to output status signals on I/O pins which can be timed using external tools. Adding instrumentation in this manner is intrusive to the application code. Not only is there a possibility of introducing bugs to the code, there is also the possibility that the instrumentation may alter the timing.

However, using the demonstration platform described it was possible to verify the timing of software executing on the CPU12X without impacting upon the application. The XGATE was used to configure the S12XDBG module to trigger the XGATE on execution of each application function. The XGATE software also included a timer interrupt routine with 1μs resolution. When triggered the XGATE software stored the execution time of the function and reconfigured the S12XDBG module for the next trigger event. In this way it was possible to measure the timing of each function without adding instrumentation software to the application code as would generally be done.

The instrumentation software was written such that the minimum and maximum execution times for each function were recorded. This summary timing data was outputted using a RS232 serial interface after each 100,000 iterations

10

of the application providing a concise record of the required data. Additionally, an execution time limit for each function was provided in the instrumentation code, enabling the XGATE to check that each function completed within the expected time and to signal if a violation occurred.

3. DISCUSSION

The demonstration platform illustrates the concept of using a secondary CPU to execute instrumentation code to extract information relating to application code residing on another CPU; however, there are a number of limitations with this platform.

The primary limitation is the lack of a direct hardware interrupt from the debug module to the XGATE. This necessitated writing a simple interrupt handler routine which resides on the CPU12X. Although this additional instrumentation code on the application CPU is minimal and isolated to a single location, it would be preferable if this could be avoided, not least because the handling of the interrupt may interfere with the execution timing of the application. However, since the XGATE is intended to function as a peripheral interrupt handler to which 108 other interrupt sources can already be routed, addition of the necessary hardware to enable direct triggering of the XGATE from the debug module should not be impractical.

The time required to execute the instrumentation code on the XGATE dictates the minimum response time between successive instrumentation triggers. This limitation does not exist when instrumentation code is added directly into the application, since such code would execute in an inherently sequential manner. This disadvantage needs to be balanced against the fact that the inline instrumentation code impacts upon the execution timing of the application, whereas instrumentation code residing on the XGATE does not. And because of this impact, inline instrumentation code is generally optimised to minimise the execution time required; the instrumentation code on the XGATE can obviously be optimised in the same way.

Instrumentation code is added to applications for a wide variety of reasons. However, given the limited bandwidth available to extract and process this data, it is common practice to filter the output data or to only enable a small portion of the instrumentation code at any given time. When a particular property is being verified, or a specific bug is being isolated, it is much preferable to only extract data related to that specific item. Therefore, in many cases the ability to only process one instrumentation trigger at a time may not be a significant limitation. If multiple or adjacent instrumentation triggers are required the XGATE architecture can support a high priority 'thread' interrupting a lower priority one. Therefore, with suitable hardware, the XGATE could potentially support more than one instrumentation trigger.

When multiple CPUs access any shared resource the potential for access conflicts and contention must be considered. The MC9S12XE architecture employs a novel interleaved RAM access scheme whereby the CPU12X and XGATE access the memory on alternate cycles; this alleviates contention issues when accessing a shared memory location. The architecture also includes eight semaphore bits to support sharing of larger data structures. In many verification and debug situations the principle aim is to monitor and record, rather than alter, data values. In these situations the ability to trigger on precise data values using the S12XDBG comparators and the ability to record data accesses to the trace memory greatly simplifies shared memory access complexities.

Despite these limitations the demonstration platform does show the feasibility of the approach proposed; and this demonstration platform is intended to act merely as a proof-of-concept rather than as a final solution.

Considering the instrumentation platform concept against the key attributes previously outlined, we can see that this approach offers many benefits. In terms of cost, the platform proposes reusing existing on-chip debug hardware and utilizing a secondary CPU core. Adding a secondary core for instrumentation purpose alone may appear to be prohibitive, but many multi-core devices include at least one CPU which is underutilized, or used for non-critical functions, which could be assigned to the instrumentation task when needed. In terms of modifications to the application software the approach proposed is minimally invasive or non-invasive. In terms of performance, the impact upon execution of the application on the primary CPU is also minimal. The ability to isolate or eliminate the instrumentation from the application code is inherent within the approach proposed. For software engineers, software instrumentation should be easy to add as no new tools or hardware are required. And the proposal does not exclude the ability to output instrumentation data in any desired format; in fact, the use of a secondary CPU to handle communications alleviates many concerns with using traditional communication ports.

Of course, a similar separation between application and instrumentation code could be achieved using a suitably partitioned OS. However, sufficient spare processing capacity would still be needed if this were not to have any performance impact.

4. CONCLUSION

Software instrumentation, requiring minimal on-chip hardware is often considered to be a simple, inexpensive, and flexible means of monitoring execution. But using software instrumentation has both advantages and disadvantages. The advantages centre upon economics and simplicity; whereas, the disadvantages centre upon the resources required and the risks posed. Considering any alteration to source code, whether application or instrumentation code, the impacts and risks are the same. An extreme approach to eliminating the disadvantages of software instrumentation is to exclude it entirely. However, that approach also eliminates all possible benefits, and as such, may be too extreme.

The author has outlined an alternative concept which isolates the instrumentation code in a secondary CPU core, which addressed most of the concerns with adding instrumentation to application code. Using software residing on the secondary core the author could configure the debug hardware to monitor execution of the software on the primary CPU. This solution also retains the key desirable attributes for software instrumentation.

5. REFERENCES

[1]   Stollon, N., *On-Chip Instrumentation: Design and Debug for Systems on Chip*, 1st ed., Springer, 2011.

[2]   Hopkins, A.B.T. and McDonald-Maier, K.D., "Debug support for complex systems on-chip: a review," *IEE Proc., Comput. Digit. Tech.*, vol. 153, no. 4, pp. 197-207, July 2006.

[3]   Vermeulen, B., Stollon, N., Kuhnis, R., Swoboda, G., and Rearick, J., "Overview of Debug Standardization Activities," *IEEE Des. Test. Comput.*, vol. 25, no. 3, pp. 258-267, May-June 2008.

[4]   Mayer, A., Siebert, H., and McDonald-Maier, K.D., "Boosting Debugging Support for Complex Systems on Chip," *IEEE Computer*, vol. 40, no. 4, pp. 76-81, April 2007.

[5]   "CoreSight Components Technical Reference Manual," ARM Limited, 2009.

[6]   "Nexus 5001 Standard for a Global Embedded Processor Debug Interface," IEEE-ISTO 5001-2003.

[7]   "White paper on MIPI Test and Debug Interface Framework," MIPI Alliance, 2006.

[8]   Brouillette, P., "Accelerating SoC Platform Software Debug with INTEL's Sven and Omar," in *Proc. System, Software, SoC and Silicon Debug Conf. (S4D)*, Southampton, UK, Sept. 2010, pp. 1-4.

[9]   Martin, A. and Rohloff, I., "Instrumentation Trace," in *Proc. System, Software, SoC and Silicon Debug Conf. (S4D)*, Southampton, UK, Sept. 2010, pp. 1-3.

[10]   Watterson, C. and Heffernan, D., "Runtime verification and monitoring of embedded systems," *IET Softw.*, vol. 1, no. 5, pp. 172-179, 2007.

[11]   Scottow, R.G., Hopkins, A.B.T., and McDonald-Maier, K.D., "Instrumentation of Real-Time Embedded Systems for Performance Analysis," in *Proc. IEEE Instrumentation and Measurement Technology Conf.*, Sorrento, Italy, April 2006, pp. 1307-1310.

[12]   Grötker, T., Holtmann, U., Keding, H., and Wloka, M., *The Developer's Guide to Debugging*, 1st ed., Springer, Netherlands, 2008.

[13]   "Road vehicles - Functional safety Part 6: Product development at the software level," ISO 26262-6:2011.

[14]   "Guidelines for the use of the C language in critical systems," MISRA-C:2004, Oct. 2004, pp. i-106.

[15]   "MC9S12XE - Family Reference Manual," Freescale, Sept. 2010.

Software Instrumentation
of
Safety Critical Embedded Systems
A Problem Statement

Mathias Ekman

Bombardier Sweden AB

721 73 Västerås, Sweden

mathias.ekman@se.transport.bombardier.com

Henrik Thane

Safety Integrity AB

Västerås, Sweden

henrik.thane@safetyintegrity.se

Abstract—The ever-increasing complexity of embedded computer software also increases the difficulty to debug and verify the correctness of the real-world execution. In addition, today the development process must often be proved to fulfill a large number of safety requirements stated in a standard or regulation. One common way to perform verification and debugging is by facilitating some method of online-monitoring, like software instrumentation. However, in safety related systems, it is not obvious that traditional software instrumentation techniques can be applied. In this article, we will elaborate on several aspects when applying software instrumentation into safety-related systems. Problems that need to be considered will be identified, but also consequences to the problems will be analyzed. (Abstract)

Keywords: software instrumentation; monitoring; safety; embedded

I. INTRODUCTION

It might not be a well know fact but many of the products we use daily have safety related software in them, controlling vital and dangerous functions. The most obvious examples are the safety related functionality of modern cars like, airbags, ABS systems, stability control systems, radar controlled cruise controls, automatic braking systems, etc. Similarly, for railways, like trains and subways, computer software controls the doors, propulsion, brakes, traffic signaling, etc. To be considered safe many of these functions have to be certified, or *free from unreasonable risk* as it is defined in one of the best practice standards (IEC61508, 2010) [1]. In some cases, legislation requires certification (e.g., the European Machine directive and the Railway directive). Certification is defined as "a formal assurance that the system has met relevant technical standards designed to ensure it will not unduly endanger the public and can be depended upon to deliver its intended

service safely and securely". Certification is also used as a unique selling point, in order to differentiate a product from the competition. Some of these standards are harmonized with the directives and regulations, meaning that if you comply with a standard you automatically comply with a regulation.

In order to prove that a computer based system complies with a given standard, the development process and the computer software must be proven to fulfill a large number of requirements stated in the standard, supported by evidence and arguments, and compiled into something called a safety-case.

Safety requirements in the IEC61508 standard are derived through a process that combines functional hazard assessment and risk analysis techniques. The aim of this process is to determine; (a) critical system functions, i.e. functions with the potential to be hazardous in the case of failure; (b) safety requirements for unavoidable hazards related to these functions, i.e. maximum tolerable failure probabilities that confine the risk associated with the operation of the system to an acceptable level; and (c) any demands for additional functions in order to achieve acceptable levels of risk for the system.

In the best practice standards, the risk of the system is defined as a relative level of risk-reduction provided by the safety case, or to specify a target level of risk reduction. The safety integrity level (SIL) determines the acceptable risk. If the risk inherent to a system exceeds the acceptable risk, the IEC61508 demands the implementation of one or more safety functions that bring down the actual risk to a tolerable level. Four levels of SIL are defined, with SIL4 being the most dependable and SIL1 being the least.

The standard also defines process measures and methods to be applied during the implementation of the safety functions. A safety-case is a logically expressed argument, supported by a sufficient body of evidence, for why a system is adequately safe to be used in a particular application. The safety argument documentation is structurally collected on top-level.

This process is usually an arduous and rigorous activity for proving that a computer system complies with a given standard, and is not especially agile.

At the same time, as many of us know, embedded computer software have a lot of bugs in them, and the ever-increasing complexity, as well as multitasking, multiprocessing distribution, and real-time aspects do not make it easier to find them. Obviously, safety related software is not exempt from these bugs, rather the opposite, if a bug leads to a failure it might lead to an accident where people get hurt or killed.

When a system is verified and debugged, it is essential and important to apply techniques and tools that increase the system-knowledge, especially for safety critical embedded systems. Developers need to have adequate understanding of the system to be able to verify the correctness of the execution. There have been several methods developed for static and dynamic code analysis. These methods analyse the source code or abstractions derived from the object code in order to observe and identify flaws in the application. The information obtained from the analysis can be used for identifying possible implementation errors, or for formal methods that mathematically prove certain properties. However, such an approach is not useful to identify timing constraints violations of real-time systems. For these cases, instrumentation techniques must be applied to increase observability and to collect run-time information, so the real-world execution behaviour of the embedded real-time system can be analysed.

An important activity to ensure a certain degree of software quality is to monitor and observe that the execution conforms to a certain safety requirement. For embedded systems, the need for observing the system becomes even higher since often only a few auxiliary interfaces are available and due to a restricted access to deployed systems.

One of the most utilized methods for debugging is based on on-line monitoring, making use of some type of technique for code instrumentation, data extraction, and post analysis; in some cases hardware based, in other cases software based.

There exist several hardware-based approaches for software observability, relying on special hardware interfaces, but they usually have limitations of scalability and availability for series production and already deployed systems. In addition, observations are on a low level, which makes the monitoring process tedious to perform [2-4].

II. SOFTWARE INSTRUMENTATION

Software instrumentation is a flexible approach for increased observability, where static or dynamic software instrumentation is used for observing execution of the system. The technique is applied by inserting additional code into an application to observe its behaviour. Instrumentation can be performed at various stages: in the source code, at compile time, post link time, or during run-time [5-8].

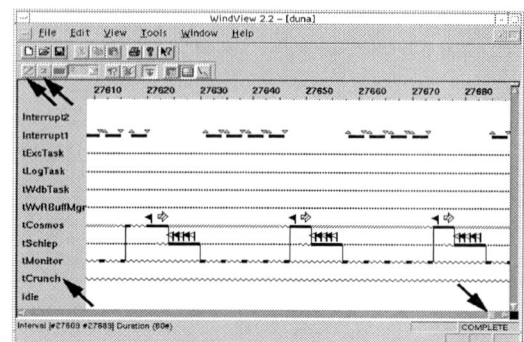

Figure 1: Software instrumentation with the WindRiver WindView tool.

Software instrumentation can be divided into two main categories: static and dynamic instrumentation. Static instrumentation relies on prepared code prior execution and is limited to only allow for activation of prepared instrumentation code. The approach required decisions to be made prior to deployment regarding selection of data collection. This approach makes it difficult to add or modify code that has been deemed necessary after deployment.

The most dynamic approach is dynamic software instrumentation, which have proven to be valuable for observing execution of applications when testing, debugging or running diagnostics on computer software. The approach allows for code to be downloaded to the target post deployment, and to activate the new instrumentation code during runtime.

Despite the fact that standards for safety certification has been adopted for quite a time, the cost of obtaining a certification is significant, estimates has been done that over 30% of the lifecycle cost are associated with the certification process [18], other studies approximates an interval of 25-75% [19]. By applying already certified techniques for software instrumentation into a development project, costs for safety certification could be significantly reduced, instead of having to perform the certification process for each change of the instrumentation software.

In a safety-case, the safety of computer software cannot be argued to be better than its weakest link. Consequently, if software instrumentation is used, the instrumentation technique chosen must be proven to be safe according to an applicable standard. In some of the standards, e.g., EN50128 (2011) [9], and IEC61508 (2010) [1], dynamic re-configuration, like in-run-time instrumentation, is not recommended for use, i.e., essentially prohibited if certification is the goal (table A2, IEC61508 [1]), unless it can be proven with evidence and arguments that it is safe to be used.

Within IEC61508, the term "highly recommended" is applied to items that must be complied with, or there must be a detailed explanation with a suitable alternative, that the assessor must accept. For support tools and programming languages, IEC61508 highly recommends:

- Use of suitable programming language
- Use of strongly-typed programming language
- Use of defined language sub-set
- Certified tools, or proven confidence from extended tool usage
- A certified translator, or proven confidence from extended translator usage
- Use of library of trusted and verified software modules and components

In order to facilitate the development of safe software instrumentation techniques, which do not introduce unreasonable risk, we state in this article a set of problems and consequences in the area that must be dealt with as a first step, before possible research solutions can be elaborated in the future.

Problem 1:
It must be proven, by giving sufficient evidence and arguments according to an applicable software safety standard:

- That the instrumentation mechanism that inserts code into the source code or executable code is safe.

Problem 2:
It must be proven, by giving sufficient evidence and arguments according to an applicable software safety standard:

- That the instrumentation code is safe, and has no adverse effect on the subject during execution.

Problem 3:
It must be proven, by giving sufficient evidence and arguments according to an applicable software safety standard:

- That the data extraction technique is safe, and has no adverse effect on the subject when data is uploaded from the target.

Problem 4:
It must be proven, by giving sufficient evidence and arguments according to an applicable software safety standard:

- That the inactivation of the instrumentation code is safe, and has no adverse effect on the subject.

Here lies the conundrum, all existing techniques known to the authors [10-15] have not addressed the above problem statements.
Consequently, existing techniques cannot be used for safety related software.

There are a number of consequences to the above stated problems:

Consequences related to problem 1:
- The instrumentation mechanism that inserts code into the source code or executable code must be developed according to some applicable safety standard, and with the rigor needed; SIL1-4 for IEC61508, ASIL A-D for ISO26262 [1, 16] .
- All functional requirements for a generic software instrumentation component must be deemed as safety related, when it is developed out-of-context, since specific hazards and related functional requirements are firsts apparent when used in a real system, in a real environment. Consequently, all requirements must be tested with the rigor required by the standard. In some cases this would lead to an unreasonable long testing time. As a result, software implementations must have such low complexity that full black-box test coverage can be achieved in a reasonable time.

Consequences related to problem 2:
- Alternative 1. The instrumentation code itself (source code or executable code), must be developed according to some applicable safety standard, and with the rigor needed (SIL1-4 for IEC61508, ASIL A-D for ISO26262) [1, 16].
- Alternative 2. Some certified infrastructure support guarantees that he instrumentation code cannot read or write to hazardous data, or in other ways impact the data, control flow, or execution time of the system in hazardous ways. Something similar to sandboxing should be applied to the instrumentation.

Consequences related to problem 3:
- The technique used for uploading data from the subject must be developed according to some applicable safety standard, and with the rigor needed (SIL1-4 for IEC61508, ASIL A-D for ISO26262) [1, 16].
- The uploading of data from the target must be proven to have no impact on safety with respect to timing, and data bandwidth.

Consequences related to problem 4:
- Essentially, this means that the instrumentation code must have no probe-effect [17], or a predictable probe-effect, which can be proven to have no safety impact before instrumentation inactivation.

III. CONCLUSION

In this paper we have identified four essential problems that must be dealt with if software instrumentation techniques are going to be used for monitoring and debugging of safety related software that has to comply with best practice or regulation standards like IEC61508, EN50128, and ISO26262, or similar. We have also elaborated on consequences that have to be considered when developing solutions to the listed problems. Since safety related software is an increasing niche, with everyday impact, researchers and developers in the area of monitoring, and debugging should consider dealing with the listed problems.

REFERENCES

1. *IEC61508. Functional safety of electrical/electronic/programmable electronic safety-related systems. International Electrotechnical Comission*, 2010.

2. Wetterquist, K., *Implementing a Boundary Scan Methodology.* EE: Evaluation Engineering, 2009. **48**(10): p. 20-24.

3. Meek, P. *68HC12- Software Debugging via Backround Debug Mode (BDM).* 1998. Miller Freeman.

4. Iyenghar, P., et al., *A model based approach for debugging embedded systems in real-time.* International Conference on Compilers, Architecture & Synthesis for Embedded Systems, 2010: p. 69.

5. Evans, D. and D. Larochelle, *Improving Security Using Extensible Lightweight Static Analysis.* IEEE Software, 2002. **19**(1): p. 42.

6. *EEMBC Adopts DoubleCheck(TM) for Its Industry-Standard Processor; Benchmarks; Green Hills Software's Static Analysis Tool Increases Code Quality*, 2007.

7. Andrew, R.D. and O.H. Jason, *nAIT: A source analysis and instrumentation framework for nesC.* The Journal of Systems & Software, 2009. **82**: p. 1057-1072.

8. Hazelwood, K. and A. Klauser. *A Dynamic Binary Instrumentation Engine for the ARM Architecture.* 2006. ACM Press.

9. *EN50128. Railway applications. Communication, signalling and processing systems. Software for railway control and protection systems*, 2011.

10. Miller, B.P., et al., *The Paradyn Parallel Performance Measurement Tools*, 1995.

11. Hollingsworth, J. and B. Buck, *An api for runtime code patching.* The international journal of high performance computing applications, 2000. **14**(4): p. 317.

12. Srivastava, A. and A. Eustace, *ATOM: A System for Building Customized Program Analysis Tools.* ACM SIGPLAN NOTICES, 2004. **39**: p. 528-539.

13. Romer, T., et al., *Instrumentation and optimization of Win32/Intel executables using etch*, in *The USENIX Windows NT Workshop: August 11-13, 1997, Seattle, Washington. USENIX B2* 1997. p. 1-7.

14. Engler, D.R., W.C. Hsieh, and M.F. Kaashoek, *C: a language for high-level, efficient, and machine-independent dynamic code generation.* Annual Symposium on Principles of Programming Languages, 1996: p. 131.

15. Hollingsworth, J.K., B.P. Miller, and J. Cargille, *Dynamic Program Instrumentation for Scalable Performance Tools*, in *Proc. Scalable High-Performance Computing Conf. B2 - Proc. Scalable High-Performance Computing Conf.* 1994: Knoxville, Term.

16. *ISO26262. Road vehicles - Functional safety*, 2011.

17. Gait, J., *A probe effect in concurrent programs*, 1986: In software-practice and experiences, volume 16(3). p. 225-233.

18. Cleaveland, R., *Formal Certification of Aerospace Embedded Software.* National Workshop on Aviation Software Systems: Design for Certifiably Dependable Systems, 2006.

19. Storey N., *Safety-Critical Computer Systems.* Addison-Wesley, 1996.

Vienna, Austria – September 19-20, 2012

Session 2: Profiling, Runtime Verification, Reverse Debug and Speedpath Diagnosis

Profiling Bare-Metal Cores in AMP Systems
Adriaan Schmidt (Fraunhofer ESK)

A Runtime Verification Unit for Microcontrollers
Thomas Reinbacher, Andreas Steininger (Vienna University of Technology),
and Martin Horauer (University of Applied Sciences Technikum Wien)

A Review of Reverse Debugging
Jakob Engblom (Wind River)

Application of Timing Variation Modeling to Speedpath Diagnosis
Mehdi Dehbashi and Görschwin Fey (University of Bremen)

www.ecsi.org/s4d

Profiling Bare-Metal Cores in AMP Systems

Adriaan Schmidt

Fraunhofer Institute for Communication Systems ESK

Munich, Germany

Email: adriaan.schmidt@esk.fraunhofer.de

Abstract—**In this paper we describe OProfileBM, an extension to the sampling-based Linux system profiler OProfile. It enables cheap and easy to use performance analysis on multi-core platforms operated using asymmetric multiprocessing (AMP). OProfileBM is linked to standalone applications that run on bare-metal cores without an operating system, and collects samples in the background, while the application is running. The collected data is then transferred to an OProfile instance running on Linux for further analysis. OProfileBM integrates seamlessly with the existing OProfile frontend.**

I. Introduction

In the optimization of software, performance analysis is an important step. Profilers represent a class of tools that are often used to gather information on the dynamic behavior of applications. They are used to learn where execution time is spent, and thus help in identifying critical locations in the program code.

In multi-core software architectures, especially on embedded systems, asymmetric multiprocessing (AMP) is an increasingly used paradigm, for example in telecommunication, where we can find integrated control and data plane implementations that utilize an AMP architecture [1]. This means that not all cores of a multi-core system are controlled by the same operating system instance, as they would be in symmetric multiprocessing. A typical software architecture could feature one or more cores running a standard operating system, while other cores are operated "bare-metal", i. e. with only a minimal runtime environment. Examples of such environments are Cavium's "Simple Executive" [2] or Freescale's "Light Weight Executive".

Applications executing in such a bare-metal environment have exclusive access to their CPU core and are not interrupted by an operating system scheduler. This makes it possible to implement applications that require a small system overhead to perform at high throughput or with low latencies. On the downside, the absence of an operating system infrastructure usually means that many profiling tools are not available.

We present in this paper a cheap, easy to use solution to collect performance data of stand-alone applications in AMP systems that run Linux on at least one of their cores. To achieve this, we have developed OProfileBM, which extends the widely used Linux system profiler OProfile [3] to bare-metal cores.

Similar tools include the "Bare Metal Performance Tools" by ENEA [4], currently available for MIPS-based platforms. An alternative option to obtain an application profile is the use of a hardware-based tracing solution [5]. While this approach has the advantage of having a smaller influence on the run time behavior of the target, it is more costly, as it requires hardware support within the CPU and a dedicated hardware device to collect the execution traces.

The next section gives a brief overview of the features of OProfile. Sections III and IV describe the concept of our extension to the profiler, and the implementation and evaluation of a prototype. Section V concludes this paper.

II. OProfile

OProfile is a system profiler that is available on most Linux platforms. It operates sampling-based, which means that the profile is obtained through a statistical analysis of random samples that allow conclusions about the run-time behavior of the system. The alternative would be instrumentation-based profiling, e. g. using the GNU profiler gprof [6], in which the compiler is instructed to insert instrumentation code into the application. In this case the application itself measures execution times of functions and records the profile. Compared to instrumentation-based approaches, which only provide information on a single application, OProfile has the advantage that it can analyze all applications in a Linux system, including shared libraries, as well as the Linux kernel and its modules.

OProfile records the value of the CPU program counter at regular intervals, thus obtaining a sample representing which code is executed during which fraction of execution time. Sampling the program counter yields source code addresses, which are processed by OProfile to represent useful information to the developer. The addresses are resolved and related to executable programs and specific lines within their sourcecode.

The sampling process of OProfile is triggered using the hardware performance counters provided by modern CPU cores. In this way it is possible to analyze different aspects of program execution that go beyond mere execution time, e. g. which functions are responsible for cache misses.

OProfile consists of a kernel module (`oprofile.ko`), a user space daemon (`oprofiled`), and a command line frontend of several tools (`opcontrol`, `opreport`, `opannotate`). The kernel module is responsible for collecting samples, and determining the address space and application binary to which the samples refer. This information is then transferred to user space, where the daemon stores it to the file system. The stored data can then be analyzed, but also archived or even transferred to a different machine for later analysis.

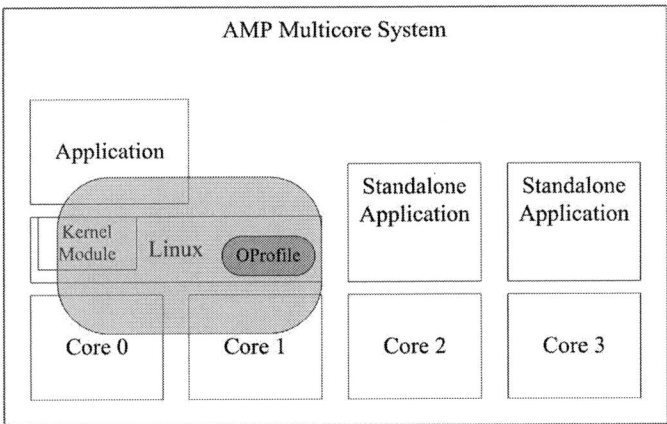

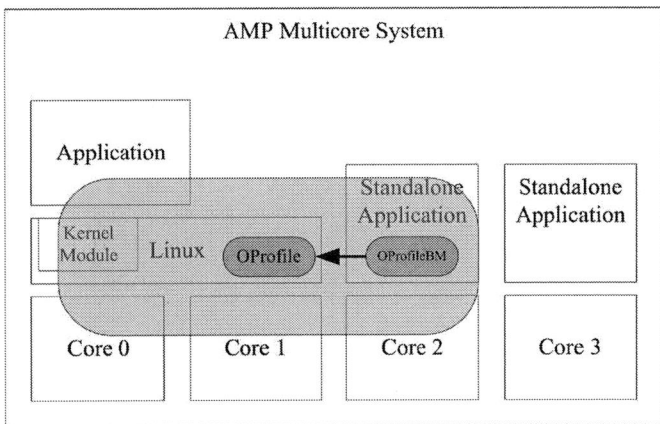

(a) Standard OProfile permits the analysis of the Linux kernel, its modules, and applications running in the Linux environment.

(b) Using OProfileBM, analysis of standalone applications becomes possible.

Figure 1. Illustration of OProfileBM

When applied to an AMP scenario, the capabilities of OProfile are limited. It can only collect data from its Linux instance and the processes running therein. Applications that run on different cores, either using other operating systems or as standalone applications, are not visible.

III. EXTENSION TO STANDALONE APPLICATIONS

We have developed OProfileBM, an extension to OProfile that allows the profiling of standalone applications running on bare-metal cores in an AMP environment. Figure 1 illustrates this, with Figure 1a showing the components that can be profiled without OProfileBM. Figure 1b depicts the visibility of system components when using OProfileBM.

One objective in designing OProfileBM was to reuse as many features and components of OProfile as possible, and not to change the way in which the user interacts with the profiler.

Figure 2 shows the main components involved in the operation of OProfileBM. Highlighted in red are the components that were added or changed to support OProfileBM.

A. Bare-Metal Runtime Module

On the bare-metal core, OProfileBM itself takes the form of a runtime module that is linked to the standalone application. Once initialized, the runtime module operates independently and has no explicit interaction with the standalone application. There can be several instances of the runtime module present in a system. This makes it possible to simultaneously analyze multiple bare-metal cores, even if they execute different standalone applications.

The sample recording of OProfileBM is triggered by interrupts that are generated by the Performance Monitoring Unit (PMU) of the CPU. Two parameters influence the behavior: the *event* specifies what is counted by the performance counters, e.g. clock cycles or cache misses, and the *reload value* determines the frequency of interrupts, e.g. every 100,000th event occurrence. These are the same parameters also used by OProfile, and which can be set by the user by means of the `opcontrol` command on Linux. The configuration made here

is transferred to OProfileBM, ensuring that sample collection on all cores is executed with the same settings.

When collecting samples on Linux, OProfile performs significant post-processing on the collected program counter addresses. The kernel and each user space application have their own address space, and to relate program counter samples to application binaries, loaded executables, libraries and kernel modules have to be resolved. In a bare-metal environment this process can be simplified. Here we only deal with one address space and one executable image. This means that no post-processing of the collected addresses is required.

The samples collected by OProfileBM are stored in a ring buffer, which is placed in a shared memory region. From here they are collected by OProfile at regular intervals. Each bare-metal core has its own private ring buffer.

Only minor changes to the bare-metal application are necessary to incorporate OProfileBM. Before profiling is possible, the runtime module needs to be initialized. The initialization function, which needs to be called explicitly by the standalone application, initializes the interrupts and sets up shared memory communication. In case the environment supports termination of the standalone application, a de-initialization function needs to be called before the application is shut down.

The operation of OProfileBM is interrupt-triggered, which means that during its initialization, interrupt handlers need to be registered. Depending on hardware platform and software architecture, this may require some additional target-specific modifications of the standalone application. It may for example be necessary to de-multiplex interrupts in software in case multiple interrupt sources share one handler.

B. Changes to OProfile

To integrate samples collected by OProfileBM into the processing and analysis of OProfile, some changes were necessary to the OProfile kernel module. During its operation, OProfile collects samples on all cores running Linux, and saves them to core-private buffers. These samples are periodically transferred to the global Event Buffer, from where they are

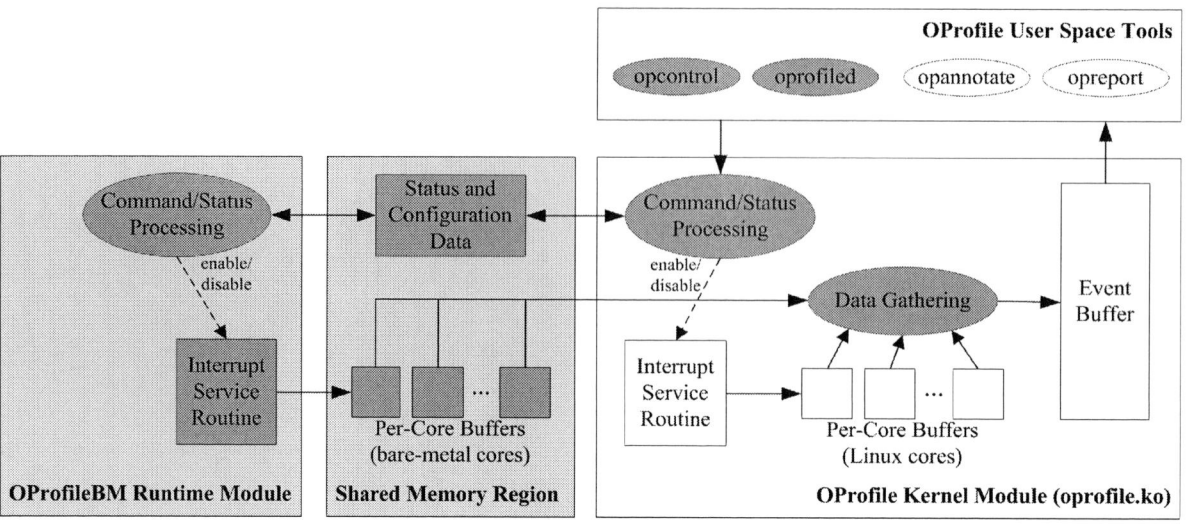

Figure 2. Relevant architecture aspects of OProfile, OProfileBM, and their communication using shared memory

read and further processed by the user space components. The internal function which periodically collects samples from the core-private buffers and writes them to the Event Buffer was extended. It now also collects samples from all registered bare-metal cores.

Internally, OProfile identifies each core by its number. This number is assigned by Linux and does not neccessarily correspond with the physical core number. OProfileBM is unaware of Linux core numberings and can only use physical core numbers to identify its instances. To avoid possible naming conflicts and to clearly identify bare-metal cores, we chose to add an offset of 1000 to the physical core numbers running instances of OProfileBM. Thus, in the OProfile frontend we might see core numbers of $0, 1, \ldots$ identifying Linux cores, and core numbers of $1000, 1001, \ldots$ identifying bare-metal cores.

With regard to the user interface, only one new parameter was introduced to `oprofiled` and `opcontrol`. With its help one specifies the image files of the standalone applications that are being executed. After providing this information, the data collected on the bare-metal cores integrates seamlessly with existing reports and analyses.

C. Interaction between OProfile and OProfileBM

For communication between OProfile running on Linux and the OProfileBM runtime module, a shared memory region is used. In addition to the per-core sample buffers already mentioned above, the shared memory holds status and configuration data for each bare-metal core. These memory locations are used to relay commands from OProfile to the OProfileBM instances. This includes instructions to start and stop sample recording, and *event* and *reload value* settings for the performance counters.

Upon initialization and deinitialization the OProfileBM runtime-modules indicate their presence in a status word. This makes it possible for the OProfile kernel module to auto-detect running instances of OProfileBM.

IV. EVALUATION AND DISCUSSION

We implemented a prototype of OProfileBM for the Cavium Octeon processor family and Cavium's Simple Executive environment. The runtime module that is running in the background of the standalone applications on the bare-metal cores has a code size of is 2428 bytes[1] and uses only few resources of the Octeon platform. For the registering of interrupt handlers and the allocation of the shared memory region we use functions provided by the Simple Executive environment. When sample collection should be started or stopped, a corresponding command is put into the configuration fields within the shared memory region. The presence of a new command is indicated to the runtime modules by issuing a core-to-core interrupt.

Recording samples of the program counter causes a runtime overhead, slowing down execution of the target application. However, because of the nature of sampling-based profiling, the overhead can be influenced. By collecting samples at lower frequencies, the overhead can be arbitrarily small. Care has to be taken that enough samples are present to still provide a representative profile. This means that if samples are collected at lower frequencies, measurements typically have to be run for longer durations to obtain the same profile accuracy.

To quantify the overhead caused by OProfileBM, we ran several different benchmark applications in the Cavium Simple Executive environment: *jpeg* encodes a 1024×768 pixel bitmap image (2.3 megabytes) to the jpeg format, *matmult* performs the multiplication of two 300×300 integer matrices, *fft* calculates the Fourier transform of a series of 65536 complex numbers, and *qsort* sorts a randomized array of 2,000,000 elements using the quicksort algorithm. We measured the execution time of each benchmark algorithm using the CPU cycle counter, once with profiling disabled, and with profiling enabled using several different settings.

We ran a series of experiments using the performance

---

[1] compiled using gcc version 4.3.3 from the Cavium Octeon SDK 2.1 and an optimization setting of -O2

| | JPEG | | MATMULT | | FFT | | QSORT | |
reload value	samples/s	slowdown	samples/s	slowdown	samples/s	slowdown	samples/s	slowdown
10000	77884	4.14%	70394	6.92%	77185	4.55%	74778	4.88%
20000	39562	2.20%	36255	3.44%	39529	2.27%	38117	2.56%
50000	15959	1.01%	14750	1.40%	15973	0.97%	15408	1.22%
100000	8007	0.57%	7421	0.71%	8022	0.47%	7750	0.69%
200000	4012	0.31%	3722	0.36%	4020	0.22%	3881	0.38%
500000	1607	0.13%	1492	0.14%	1610	0.08%	1555	0.18%
1000000	804	0.07%	746	0.08%	806	0.03%	780	0.10%

Table I

EFFECT OF OPROFILEBM ON THE RUN-TIME BEHAVIOR OF SELECTED BENCHMARK APPLICATIONS

counters to count several different events and using different reload values. From the measured execution times we calculated the slowdown incurred when using OProfileBM. In addition we count the mean number of samples collected per second. Table I presents the measurement results when using the number of clock cycles as event counted by the performance counter. The clock cycle event is the event used by OProfile by default, and probably the one most commonly used in practice, as it yields a profile of how much execution time was spent in which parts of the application. The default reload value is 100,000.

We can see that the overhead becomes smaller if samples are collected at a lower frequency. Figure 3 confirms this, and reveals a generally linear behavior. In addition it can be seen that the sample collection has different impact on our different benchmarks. This is consistent with results from [7] or [8] who found that interference caused by shared caches does not affect all applications equally.

V. CONCLUSION AND FUTURE WORK

We presented OProfileBM as an extension of the statistical profiler OProfile to bare-metal environments. It allows for the profiling of standalone applications in AMP multi-core systems. Benefits of OProfileBM include:

- Cheap and easy way to obtain performance data that is otherwise not accessible
- Low overhead during profiling; no overhead while profiling is turned off
- Only minimal changes neccessary to standalone application (OProfileBM init function needs to be called)
- Easy to use with the existing OProfile user interface

In the future we plan to continue our work on performance analysis in bare-metal environments. With Linux perf [9] there is a new set of tools capable of performance measurement and profiling. Its suitability for bare-metal applications is so far unexplored. To provide more detailed information on the execution behavior of standalone applications, it may also be useful to adapt solutions for instrumentation-based tracing to bare-metal environments, e. g. LTTng [10].

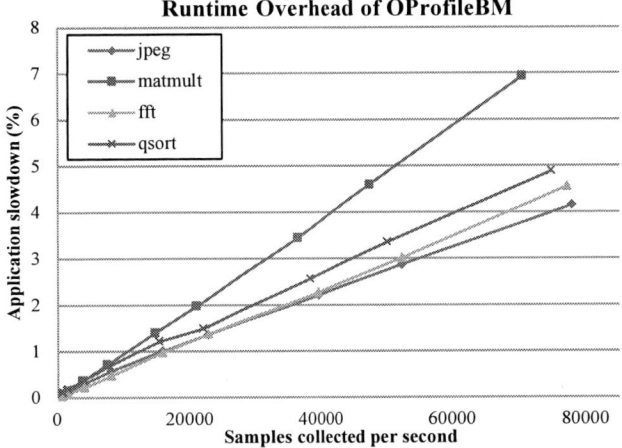

Figure 3. Application slowdown as a function of the number of samples collected per second

REFERENCES

[1] P. Strömblad, *ENEA Multicore: High performance packet processing enabled with a hybrid SMP/AMP OS technology*, white paper, 2009

[2] Cavium Networks, *OCTEON Programmer's Guide, The Fundamentals, Introduction Version for Cavium Networks University Program*, July 2010

[3] J. Levon and P. Elie, *OProfile: A System Profiler For Linux*, http://oprofile.sourceforge.net, 2004

[4] ENEA AB, *ENEA Bare Metal Performance Tools for Netlogic XLP and Cavium Octeon Plus*, data sheet, 2011

[5] M. Lindahl, *Using Hardware Trace for Performance Analysis*, blog post, http://drdobbs.com/tools/184406289, October 2005, retrieved June 2012

[6] J. Fenlason and R. Stallman. *GNU prof - The GNU Profiler*, Free Software Foundation, Inc., 1997. http://sourceware.org/binutils/docs/gprof/

[7] D. Eklov, D. Black-Schaffer, and E. Hagersten, *Fast Modeling of Shared Caches in Multicore Systems*, Proceedings of the 6th International Conference on High Performance and Embedded Architectures and Compilers, HiPEAC 2011

[8] D. Chandra, F. Guo, S. Kim, and Y. Solihin, *Predicting Inter-Thread Cache Contention on a Chip Multi-Processor Architecture*, Proceedings of the 11th International Symposium on High-Performance Computer Architecture HPCA-11, 2005

[9] Arnaldo Melo. *The new linux 'perf' tools*, in 17 International Linux System Technology Conference (Linux Kongress), Nuremberg, Germany, 2010

[10] M. Desnoyers and M. Dagenais, *The LTTng tracer: A low impact performance and behavior monitor for GNU/Linux*, Proceedings of Linux Symposium, Ottawa, Canada, 2006

A Runtime Verification Unit for Microcontrollers

Thomas Reinbacher[1], Martin Horauer[2], Andreas Steininger[1]

[1] Embedded Computing Systems Group, Vienna University of Technology
Treitlstraße 3, 1040 Vienna, AT
{reinbacher, steininger}@ecs.tuwien.ac.at
[2] Department of Embedded Systems, University of Applied Sciences Technikum Wien
Höchstädtplatz 5, 1200 Vienna, AT
horauer@technikum-wien.at

Abstract—**In this paper, we advocate the idea of a runtime verification unit that complements the standard concept of a watchdog unit commonly employed on a microcontroller. Such a unit provides ways to trigger counter-measures by more expressive conditions than typical watchdog conditions, such as propositions stated in temporal logics. We show a possible application of this unit in a case study where we apply our approach to a real-life example.**

I. INTRODUCTION

Embedded systems are an important enabler for 'intelligent' products and are becoming prevalent in our daily lives. Consequently, safety, security, reliability, and power efficiency aspects are of utmost importance; this is especially true when used in harsh (e.g., automotive, aerospace) or sensitive (e.g., medical) environments.

The development of respective solutions, however, struggles with ever increasing complexity that needs to be mastered in shorter development cycles. To tackle the consequential design challenge various approaches are used, e.g., reuse of proven components or moving towards higher levels of abstractions. The increasing system complexity not only makes verification more desirable, it also makes it more difficult: Our ability to design something often outperforms the ability to verify it, a phenomenon described by the *design verification gap* in the design automation community. One specific problem of embedded systems (in contrast to general-purpose computers) is their tight interaction with the environment via various sensors and actuators thus – when using an event-triggered approach – potentially interrupting the embedded system at arbitrary states. This in turn drastically increases the state space (i.e., the number of possible states the system may reside in) one needs to cope with during verification. The range of tools and methodologies to tackle the verification problem ranges from testing to formal methods (e.g., model-checking, theorem proving) all with their pros and cons. Nevertheless, systems eventually fail due to errors introduced, e.g., by the toolchain, bugs that go unnoticed, or due to external effects (e.g., EMC or natural radiation problems), see [8]. These issues are all hard to assess at the verification stage [2], [32]. A common way to deal with rare events is by using watchdog timers. A watchdog timer serves as a monitoring component in a wide range of applications. The task of a watchdog is to wait for regular service from the component being monitored. In case the service discontinues the watchdog may perform a *counter-measure*, e.g., initiate a system-wide reset in hope of re-establishing the correct operation of the system. *Watchdog processors* provide a somewhat more sophisticated check, most often of the control flow [19], [27].

In this paper, we advocate the idea of a *runtime verification unit* (RVU) that complements the standard concept of a watchdog. Such a unit provides ways to trigger a *counter-measure* by more expressive conditions, such as propositions qualified in terms of temporal logic enriched with time bounds. To illustrate, a proposition could read like: *In every possible run of the system, whenever an acknowledge is sent, we must have seen a preceding request*. A counter-measure, e.g., could be a hard reset of the unit monitored, however, when short recovery time is an issue, a more fine grained action is required and may include restarting individual tasks or a fall-back to a redundant spare part.

II. RUNTIME VERIFICATION UNIT

In recent years, runtime verification has gained increasing interest as it aims at bridging the design verification gap by connecting the traditional fields of formal verification and testing [3], [20]. In runtime verification one aims to answer the following question: Given a program Π, an execution trace π of Π and a specification ϕ, does π satisfy ϕ? As in testing, runtime verification does not prove the absence of errors, yet it provides a systematic way to link a single execution to a formal specification. In contrast to testing the execution paths are not artificially created, but rather the traces that actually occur during operation are used. Test oracles, which reflect the specification, are either automatically derived (e.g., from a functional specification given in some formal notation (logic)) or formulated manually in some form of executable code. Correctness of an execution is then judged by means of a monitor which evaluates sequences of events in an instrumented version of the program under scrutiny. Traditionally, runtime verification targets high-level software implementations. Herein, monitors are either deeply embedded in the software (instrumented) or executed as separate tasks when a real-time operating system is used. As a side-effect these

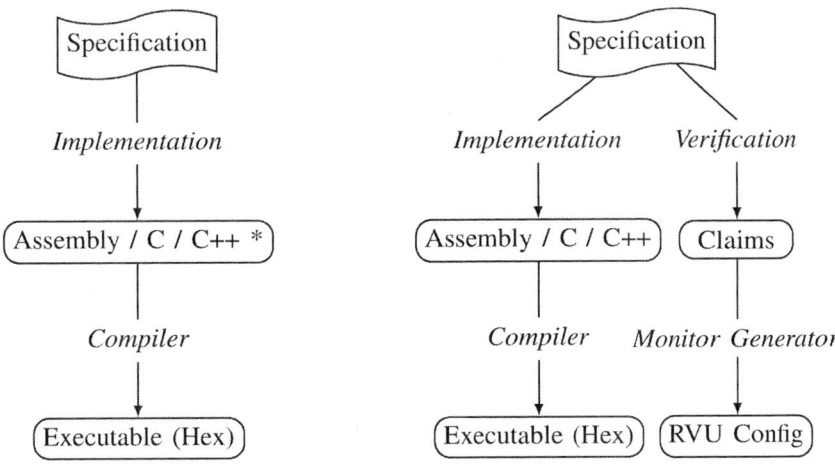

Fig. 1. Traditional (left) vs. proposed workflow (right).

approaches are rather intrusive, may impair on the functionality, and presume a correct, error free toolchain.

A. Workflow

Our approach is non-intrusive and since the workflow is parallel to the 'standard' toolchain the monitoring may not be affected by flaws of the latter. Our RVU allows monitoring embedded systems software on-the-fly in hardware, implemented as a peripheral unit for a COTS microcontroller. Figure 1 illustrates the proposed workflow. In the traditional workflow, sanity checks are typically directly infused into the high level source code, for example, an additional supervisor task which monitors the correct operation of the software system. However, for small-sized embedded target platforms the resulting additional load might be problematic due to various reasons:

- Code instrumentation increases memory consumption. This factor is of economical relevance for small-sized embedded target platforms and applies to both, the program memory as well as RAM. The relevance of this aspect is emphasized by the fact that in critical environments [10] (such as nuclear power plants) restricted architectures are often used to avoid unpredictable timing behavior, e.g., caused by caches.
- The timing behavior of the system is altered by code instrumentation [7], [12]. The additional runtime overhead may have devastating consequences for heavy-loaded real-time applications with tight deadlines.
- Code instrumentation alters the program and is thus impracticable for certified systems as it requires the re-certification of the affected program parts.

We consider implementation and verification (at run time) as two separate tasks. For the verification part, the test engineer derives correctness claims (e.g., a set of temporal formulae) based on the given specification. Next, the respective claims are used by a generator to provide the required configuration for our RVU that monitors these claims during execution of the microcontroller program. In this way, we do not alter the code base of the system under scrutiny.

B. Runtime Verification Unit

Figure 3 illustrates the structure of the runtime verification unit prototype implemented in a FPGA. It basically hosts a JTAG programming interface to configure the unit, allows monitoring the data, control (and optional address) interfaces, and provides a lean interface to the target CPU. A possible application for this runtime verification unit is as a replacement for a traditional watchdog unit on a microcontroller, as shown in Figure 2.

The RVU itself basically consists of: a) an *assertion checker* that allows the evaluation of assertions over program variables, b) a simple, programmable 'verification engine' that allows to determine the satisfaction of the specified correctness claims based on the results of the assertion checker, and c) a real-time clock unit that is required to evaluate time constraints.

(a) The *Assertion checker* supports claims that consist of conjunctions of linear constraints, where each constraint ranges over two program variables (or memory locations) only. In addition, each variable can be negated and multiplied by a constant value that is a power of two. More formally, the supported class of constraints consists of formulas, such as:

$$(\pm 2^n \cdot v_1 \pm 2^m \cdot v_2) \bowtie C$$

where v_1 and v_2 are program variables, $C, n, m \in \mathbb{Z}$, and $\bowtie \in \{=, \neq, \leq, \geq, <, >\}$ is a relational operator. The second operand v_2 is optional, thus allowing for interval constraints of the form $\pm 2^n \cdot v_1 \bowtie C$. Informally, one can monitor, whether the values of two given variables stay within defined ranges. Note that in embedded systems, such a condition could be trivially missed, e.g., when it takes multiple instructions to modify a variable and the operation gets intercepted by an interrupt where the respective service routine also accesses the variable, see [24], [26] for some examples.

(b) The *Verification Engine* is a RISC-type microprocessor featuring an instruction set that supports sequential evaluation

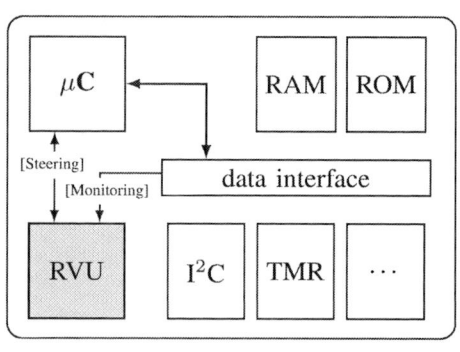

Fig. 2.   Typical Application.

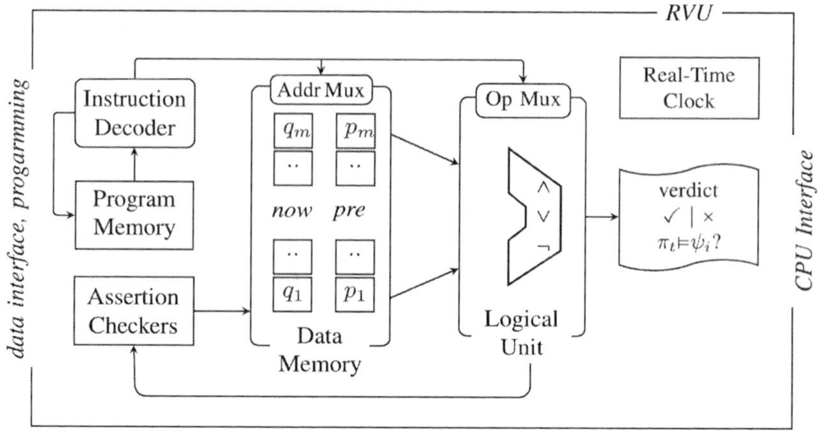

Fig. 3.   Runtime Verification Unit.

of claims provided in past-time linear temporal logic. It has separate address spaces for program and data memory (Harvard architecture). The data memory consists of two registers: one holding the evaluations of all m subformulae of ψ in the current execution cycle $q_{1...m}$ and one holding the results from the previous cycle $p_{1...m}$. With this module and the algorithms described in [14] one can check the temporal correctness claims with respect to an execution of the microcontroller, i.e., a single path of the deployed embedded software.

(c) The *Real-Time Clock* unit, finally, is used to keep track of time, timestamp events, and is used to monitor the temporal conditions of the provided claims. While a conventional watchdog has a single timeout only, our RVU allows handling various timing conditions in parallel. Upper and lower bounds on intervals can be checked, whose start and end events are derived from the conjunctions above. Our implementation is based on a global time base and thus scales favourably with the number of conditions and the interval length.

III. Specifications

The task in formal verification is to determine whether or not a given specification holds on a model, which is derived from the system under scrutiny. However, to automate this verification task, the specification language shall have the following properties:

- Unambiguous. The syntax and semantics of the specification language are well-defined.
- Expressive. Typical properties are easily expressed within the specification language.
- Easy to use. The specification language is easy to understand and comprehend, even without prior experience with formal logics.

For runtime verification the specification language also needs to allow to be synthesized into low footprint monitors. These monitors can be of different form, for example: hardware blocks, software programs, and state machines. In recent years several specification formalisms emerged. A particularly popular one is the past time fragment of linear temporal logic (LTL) commonly

referred to as ptLTL. Its popularity is mainly due to the following facts:

- Monitor generation for a ptLTL formula is straightforward, as shown in [13, Sect. 5].
- ptLTL can easily express typical specifications, as argued in [15], [17].
- Satisfaction of any ptLTL formula can be determined along the execution trace by evaluating only the current state and the results from the predecessor state.

To keep the paper self contained, we will give a short introduction to ptLTL in the following.

a) Atomic Propositions: An atomic proposition $\sigma \in AP$ is a proposition whose truth value does not depend on any other proposition, i.e., is an elementary statement in the specification. In our framework, atomic propositions directly map to the linear constraints supported by the assertion checker.

b) ptLTL Syntax: With $\bullet \in \{\wedge, \vee, \rightarrow\}$, a formula ψ is defined as:

$$\psi \quad ::= \quad true \mid false \mid \sigma \mid \neg\psi \mid \psi \bullet \psi$$
$$\odot \psi \mid \diamondsuit \psi \mid \boxdot \psi \mid \psi \, S_s \, \psi \mid \psi \, S_w \, \psi$$

$\odot\psi$ means *previously* ψ, i.e., it is the past-time analogue of next. Likewise, the other temporal operators are defined as: $\diamondsuit\psi$ expresses *eventually in the past* and $\boxdot\psi$ is referred to as *always in the past*. The duals of the until and the weak until operator are S_s and S_w, i.e., *strong since* and *weak since*, respectively. These basic operators in ptLTL can be augmented with so-called *monitoring operators* [14], [16]. The semantic of the monitoring operators is derived from the set of basic operators in ptLTL. Hence, they do not add expressive power, but provide syntactic sugar. However, using the monitoring operators, a more succinct representation of the most common properties emerging in practice is obtained. Syntactically, the monitoring operators are defined as:

$$\psi \quad ::= \quad \uparrow\psi \mid \downarrow\psi \mid [\psi, \psi)_s \mid [\psi, \psi)_w$$

$\uparrow\psi$ and $\downarrow\psi$ are trigger conditions where $\uparrow\psi$ stands for *start*

24

ψ (i.e., ψ was *false* in the previous state and is *true* in the current state, which is equivalent to $\psi \wedge \neg \odot \psi$), $\downarrow \psi$ for *end* ψ (ψ was *true* in the previous state and is *false* in the current state, equivalent to $\neg\psi \wedge \odot\psi$). The interval operators are strong *interval* $[\psi_1, \psi_2)_s$ (ψ_2 was never *true* since the last time ψ_1 was *true*, including the state when ψ_1 was *true*, equivalent to $\neg\psi_2 \wedge ((\odot\neg\psi_2)\ S_s\ \psi_1)$) and weak *interval* (equivalent to $\boxdot\neg\psi_2 \vee [\psi_1, \psi_2)_s$). For a precise definition of the semantics of ptLTL, we refer to more elaborate sources [11], [13], [16].

As an example let us consider the following specification: "If we first observe r_{20} becoming equal to 10, then it must have held the value 12 in the previous state. In ptLTL the specification ψ_{tmp} is then formalized as:

$$\psi_{tmp} \quad ::= \quad \uparrow(r_{20} = 10) \rightarrow \odot(r_{20} = 12)$$

We implement a time-bounded analogue of the *always in the past* $\boxdot$ ptLTL operator into our framework [22], denoted by $\boxdot_\tau$. Informally, the semantics of the operator w.r.t. to the current timestamp n_i are given by: $\tau \in \mathbb{N}_0$ is a time bound and $\boxdot_\tau \psi$ is true whenever ψ remains true for the interval $[n_i - \tau, n_i]$. In all other cases ψ evaluates to false. To illustrate, evaluating $\boxdot_3\,\sigma_2$ on the execution trace shown in Fig. 4 at $n_i = 16$ yields true, whereas $\boxdot_6\,\sigma_2$ yields false.

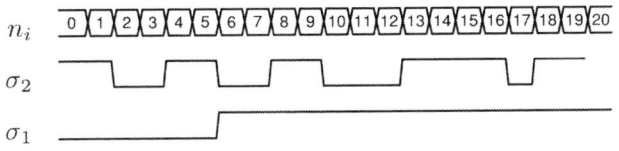

Fig. 4. Sample execution trace.

IV. CASE STUDY

To illustrate our approach we turn to an automotive case-study where we use a simplified implementation of a power-window control unit (see [23] for a more elaborate model). Fig. 5 presents the functional behavior of the respective implementation in the form of a Moore finite-state-machine that consists of the following seven states:

Q_s .... remain in the current positions (initial)
Q_c, Q_o .... fully closed, opened
Q_{um}, Q_{dm} .... being opened, closed manually
Q_{ua}, Q_{da} .... being opened, closed automatically

The inputs consist of a switch (positions u up, d down, and x not activated), an end-position sensor e, and a current value s obtained by measuring the current of the window-lift motor. Whenever an end-position is reached the window shall remain in the Q_o fully opened or Q_c closed states, respectively. Upon closing when the window hits an obstacle the current of the motor will increase. When a predefined threshold is reached s will lead to a state switch to Q_{da} (open the window automatically). Furthermore, a timer t is used to control the transition from manual to automatic mode that typically occurs whenever the switch is in the u or d position for more than a predefined time. The output controls the motor movement, i.e.,

up, down, or off.

Note, this example doesn't consider other inputs like a remote control switch (as typically available from the drivers position in a car), power supply, or inputs from a safety system.

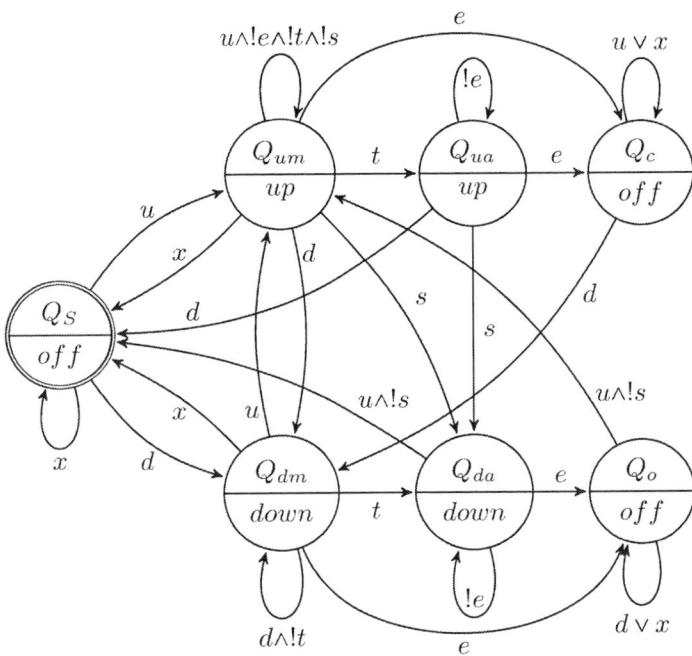

Fig. 5. Simplified power windows control FSM for one window.

Using the specification for the power-windows application we are able to derive properties in the form of ptLTL formulas. Using the latter along with the hexfile of the executable our framework is able to automatically derive a monitor that checks the execution of the embedded software implementation and upon violation generates a verdict.

A. Runtime Verification – The 'Watchdog Way'

A typical embedded application would complement the implementation of the above state-machine with a watchdog timer. The latter would trigger a reset and restart the application whenever execution of the code would fail (e.g., due to a coding error or an error introduced by the toolchain).

With the presented runtime verification unit a simple yet intuitive sanity check would, e.g, try to determine whether the state machine remains in one of the states as shown in Fig. 5 at all times. This can be expressed by the following invariant:

$$\psi_1 \quad ::= \quad \boxdot \mathtt{st} = (Q_s \vee Q_{dm} \vee Q_{da} \vee Q_{um} \vee Q_{ua} \vee Q_c \vee Q_o)$$

Another possible check could verify whether the application resides in either the opened Q_o or closed state Q_c when the end-position sensor got activated:

$$\psi_2 \quad ::= \quad \uparrow(\mathtt{st} = Q_c) \rightarrow [\uparrow e, \uparrow(\mathtt{st} \neq Q_c))_s$$
$$\psi_3 \quad ::= \quad \uparrow(\mathtt{st} = Q_o) \rightarrow [\uparrow e, \uparrow(\mathtt{st} \neq Q_o))_s$$

For either example, failing to verify one of the properties ψ_1 to ψ_3 would result in activating the output of the runtime verification unit. The latter could, e.g., transfer the state-machine to the initial state Q_s and register the fault along with information about the failing property in some error log.

B. Runtime Verification - Added Benefits

Using the runtime verification unit, however, one could easily provide added value for a system by adding checks about component failures in addition to functional execution failures as described before.

For instance, consider the case when the window is in the state Q_o, the end-positions sensor broke and remains stuck-at-one ($e = 1$), and the switch generates an u input. In this case, the state-machine would immediately switch to the closed state Q_c, despite the window being halfway closed. In this example, such an invalid state transition could be detected, when adding timing information to the verification property, i.e., the minimum closing time – the time it takes for the window to move from Q_o to Q_c under fully operational conditions. The respective property can be encoded by the following constraint (whenever the state machine transitions to state Q_c, we have not seen state Q_o for (at least) the past three seconds):

$$\psi_4 \quad := \quad \uparrow (\mathsf{st} = Q_c) \rightarrow \boxdot_\tau (\mathsf{st} \neq Q_o)) \quad | \quad \tau = 3\mathrm{s}$$

In contrast, when the finite-state-machine is in one of the states Q_{ua} or Q_{da} and the end-position sensor fails with stuck-at-zero ($e = 0$) then states Q_o or Q_c will never be reached. The motor will continue to open/close the window even though it is already at its physical end position. One scenario could be that the current threshold will be triggered and the system will end up in state Q_{da}.

$$\psi_5 \quad := \quad \neg \boxdot_\tau ((\mathsf{st} = Q_{ua}) \wedge \neg s) \quad | \quad \tau = 5\mathrm{s}$$
$$\psi_6 \quad := \quad \neg \boxdot_\tau ((\mathsf{st} = Q_{da}) \wedge \neg(u \wedge \neg s)) \quad | \quad \tau = 5\mathrm{s}$$

One more scenario could be that the timer plays havoc and fires immediately after the system reaches either Q_{um} or Q_{dm}. In this case, even though the user might have only wanted to open/close the window for a short distance, the window will eventually open up or close entirely. Once again, monitoring the timing could help identify and detect such an error.

$$\psi_7 \quad := \quad \uparrow (t \wedge (\mathsf{st} = Q_{ua})) \rightarrow \boxdot_\tau (\mathsf{st} = Q_{um}) \quad | \quad \tau = 0.5\mathrm{s}$$
$$\psi_8 \quad := \quad \uparrow (t \wedge (\mathsf{st} = Q_{da})) \rightarrow \boxdot_\tau (\mathsf{st} = Q_{dm}) \quad | \quad \tau = 0.5\mathrm{s}$$

In many similar examples runtime verification can aid in detecting respective failures along with information about the failing property, steer the system into a suitable state, and report the problem in order to trigger maintenance.

V. RELATED WORK

Watterson and Heffernan [31] give a thorough survey of approaches to monitor embedded systems in a way that is suitable for use in runtime verification methods. Hardware monitors that probe one or more internal signals have been known in literature for a few decades. An early instance thereof is the non-interference monitoring and replay mechanism by Tsai et al. [30]. Their monitoring system is based on

the MC6800 processor that records the execution history of the target system. A dedicated replay controller then replays stored executions, which supports test engineers in low-level debugging. Although we share a similar idea of probing internal signals, our framework detects specification violations on-the-fly, rather than replaying traces from some execution history.

A. Non-instrumenting Observers

The Dynamic Implementation Verification Architecture (DIVA) exploits runtime verification at intra-processor level [1]. Whenever a DIVA-based microprocessor executes an instruction, the operands and the results are sent to a checker which verifies correctness of the computation; the checker also supports fixing an erroneous operation. Chenard [6] presents a system-level approach to debugging based on in-silicon hardware checkers. While their work aims at checking functional correctness of computations, we tackle an orthogonal problem, i.e., runtime verification of embedded software. The work of Brörkens and Möller [4] is akin to ours in the sense that they also do not rely on code instrumentation to generate event sequences. Their framework, however, targets Java and connects to the bytecode using the Java Debug Interface so as to generate sequences of events.

B. Synthesis of Hardware Circuits as Observers

Lu and Forin [18] present a compiler from PSL to VERILOG, which translates a subset of PSL assertions about a software program (written in C in their approach) into hardware execution blocks for an extensible MIPS processor, thus allowing for transparent runtime verification without altering the program under investigation. The synthesized verification unit is generated by a property rewriting algorithm developed by Roşu and Havelund [25]. Atomic propositions are restricted to allow only a single comparison operator. In comparison, our approach supports more complex relations among memory values in the atomic propositions, thus yielding greater flexibility and expressiveness in the specification language. Drechsler [9] describes an approach to synthesize checkers for online verification of SoC designs, but does not allow for checking arithmetic relations among bit-vectors. For hardware designs, these specifications are often directly available from the specification [28].

A hardware-related tool called BUSMOP [20] is based on the MOP framework [5]. In essence, BUSMOP is a hardware-monitoring device which *sniffs* traffic transmitted between COTS embedded components attached to a PCI/PCI-X bus, thereby acting as *advanced* bus guardian. Similar to our approach, the monitor and the system under verification are executed within an FPGA. The specification is translated by the MOP framework into a hardware description, which is then transformed into a netlist and loaded into dynamically reconfigurable blocks of the FPGA. The framework was later adapted to different bus structures [21]. Whereas BUSMOP is designed to monitor data transmissions through a PCI (or comparable) interconnection for large-scale embedded systems, our framework targets smaller, more restricted systems.

Shobaki [29] introduced a non-intrusive framework called MAMON that monitors events on logic and system level in single and multiprocessor real-time systems. A hardware probe-unit is integrated into a hardware based real-time kernel, targeting FPGA or ASIC designs. The framework performs offline monitoring where time-stamped events are transferred to a database on a host computer. The host application operates on an event-database with the aim of providing debugging facility for a specific RTOS kernel, however, it does not support temporal logic or real-time properties.

VI. Conclusion

In this paper we advocate the idea of deploying a runtime verification unit as intelligent error detection mechanism in microcontrollers. This allows to strictly separate the tasks of implementation and verification in the embedded software design flow and circumnavigates the issue of altering the original implementation to embed runtime monitors into the system. Our monitor runs external to the microcontroller, thus, does not interfere with the original implementation. We showed how to use the past-time temporal logic ptLTL to specify safety properties emerging in the verification of a power windows application. As a next step, we want to extend this approach to more expressive logics, such as the property specification language (PSL) or metric temporal logic (MTL).

Acknowledgment

The work of Thomas Reinbacher and Andreas Steininger has been supported within the FIT-IT project CevTes managed by the Austrian Research Agency FFG under grant 825891.

References

[1] Todd M. Austin. DIVA: A reliable substrate for deep submicron microarchitecture design. In *MICRO*, pages 196–207. IEEE, 1999.

[2] G. Balakrishnan and T. W. Reps. WYSINWYX: What You See Is Not What You eXecute. *ACM Trans. Program. Lang. Syst.*, 32(6), 2010.

[3] Howard Barringer, Yliès Falcone, Bernd Finkbeiner, Klaus Havelund, Insup Lee, Gordon J. Pace, Grigore Rosu, Oleg Sokolsky, and Nikolai Tillmann, editors. *Runtime Verification - First International Conference, RV 2010, St. Julians, Malta, November 1-4, 2010. Proceedings*, volume 6418 of *Lecture Notes in Computer Science*. Springer, 2010.

[4] Mark Brörkens and Michael Möller. Dynamic event generation for runtime checking using the JDI. *Electronic Notes in Theoretical Computer Science*, 70(4):21 – 35, 2002.

[5] Feng Chen and Grigore Roşu. MOP: An efficient and generic runtime verification framework. In *OOPSLA*, pages 569–588. ACM, 2007.

[6] Jean-Samuel Chenard. *Hardware-based Temporal Logic Checkers for the Debugging of Digital Integrated Circuits*. PhD thesis, McGill University, 2011.

[7] Christian Colombo, Gordon J. Pace, and Gerardo Schneider. Safe runtime verification of real-time properties. In *FORMATS*, volume 5813 of *LNCS*, pages 103–117. Springer, 2009.

[8] Nachum Dershowitz. Software horror stories. website, accessed May 2012, http://www.cs.tau.ac.il/ nachumd/horror.html.

[9] R. Drechsler. Synthesizing checkers for on-line verification of system-on-chip designs. In *ISCAS*, volume 4, pages IV–748 – IV–751 vol.4, May 2003.

[10] Antoine Druilhe, Frederic Daumas, and Thuy Nguyen. Formal verification of an FPGA emulation of the motorola 6800 microprocessor. In *7th Int'l Topical Meeting on Nuclear Plant Instrumentation, Control and Human Machine Interface Technologies (NPIC&HMIT 2010)*, pages 1316 – 1325. American Nuclear Society, November 2010. ISBN: 978-0-89448-84-3.

[11] E. Allen Emerson. The beginning of model checking: A personal perspective. *25 Years of Model Checking: History, Achievements, Perspectives*, pages 27–45, 2008.

[12] S. Fischmeister and P. Lam. Time-aware instrumentation of embedded software. *IEEE Transactions on Industrial Informatics*, 6(4):652–663, November 2010.

[13] K. Havelund and G. Roşu. An overview of the runtime verification tool Java PathExplorer. *Form. Methods Syst. Des.*, 24(2):189–215, 2004.

[14] Klaus Havelund and Grigore Rosu. Synthesizing monitors for safety properties. In *TACAS*, LNCS, pages 342–356. Springer, 2002.

[15] François Laroussinie, Nicolas Markey, and Ph. Schnoebelen. Temporal logic with forgettable past. In *LICS*, pages 383–392. IEEE, 2002.

[16] Insup Lee, Sampath Kannan, Moonjoo Kim, Oleg Sokolsky, and Mahesh Viswanathan. Runtime assurance based on formal specifications. In *PDPTA*, pages 279–287, 1999.

[17] Orna Lichtenstein, Amir Pnueli, and Lenore Zuck. The glory of the past. In *Logics of Programs*, volume 193 of *LNCS*, pages 196–218. Springer, 1985.

[18] Hong Lu and Alessandro Forin. The design and implementation of P2V, an architecture for zero-overhead online verification of software programs. Technical Report MSR-TR-2007-99, Microsoft Research, 2007.

[19] A. Mahmood and E.J. McCluskey. Concurrent error detection using watchdog processors-a survey. *Transactions on Computers*, 37(2):160 –174, 1988.

[20] Rodolfo Pellizzoni, Patrick Meredith, Marco Caccamo, and Grigore Rosu. Hardware runtime monitoring for dependable COTS-based real-time embedded systems. *Real-Time Systems Symposium*, pages 481–491, 2008.

[21] Rodolfo Pellizzoni, Patrick Meredith, Min-Young Nam, Mu Sun, Marco Caccamo, and Lui Sha. Handling mixed-criticality in soc-based real-time embedded systems. In *EMSOFT*, pages 235–244, New York, NY, USA, 2009. ACM.

[22] T. Reinbacher, M. Függer, and J. Brauer. Real-time runtime verification on chip. In *RV*, Lecture Notes in Computer Science. Springer, 2012. accepted.

[23] T. Reinbacher, J. Geist, P. Moosbrugger, M. Horauer, and A. Steininger. Parallel runtime verification of temporal properties for embedded software. In *MESA*, pages 224–231, 2012.

[24] T. Reinbacher, M. Kramer, M. Horauer, and B. Schlich. Motivating model checking for embedded systems software. In *Mechatronic and Embedded Systems and Applications (MESA 2008)*, pages 546–551, 2008.

[25] Grigore Roşu and Klaus Havelund. Rewriting-based techniques for runtime verification. *Automated Software Eng.*, 12(2):151–197, 2005.

[26] B. Schlich and S. Kowalewski. Model checking C source code for embedded systems. In *Proceedings of the IEEE/NASA Workshop Leveraging Applications of Formal Methods, Verification, and Validation (ISoLA 2005)*, 2005.

[27] M.A. Schuette and J.P. Shen. Processor control flow monitoring using signatured instruction streams. *Transactions on Computers*, C-36(3):264 –276, 1987.

[28] Kanna Shimizu, David L. Dill, and Alan J. Hu. Monitor-based formal specification of pci. In *Proceedings of the Third International Conference on Formal Methods in Computer-Aided Design*, FMCAD '00, pages 335–353, London, UK, 2000. Springer-Verlag.

[29] Mohammed El Shobaki. On-chip monitoring of single- and multiprocessor hardware real-time operating systems. In *Proceedings of the 8th International Conference on Real-Time Computing Systems and Applications (RTCSA)*, March 2002.

[30] J. J. P. Tsai, K. Y. Fang, H. Y. Chen, and Y.D. Bi. A noninterference monitoring and replay mechanism for real-time software testing and debugging. *IEEE Trans. Softw. Eng.*, 16:897–916, 1990.

[31] C. Watterson and Donal Heffernan. Runtime verification and monitoring of embedded systems. *IET Software*, 1(5):172–179, 2007.

[32] X. Yang, Y. Chen, E. Eide, and J. Regehr. Finding and Understanding Bugs in C Compilers. In *PLDI*, pages 283–294. ACM Press, 2011.

A REVIEW OF REVERSE DEBUGGING

Jakob Engblom

Wind River Systems

ABSTRACT

Reverse debugging is the ability of a debugger to stop *after* a failure in a program has been observed and *go back* into the history of the execution to uncover the reason for the failure.

Long the dream of programmers, over the past decade, reverse execution has become a practical technique available in a number of free and commercial tools.

This article will review the history and techniques of reverse debugging, as researched, implemented, and used from the 1970s until today. We will provide some personal insights into reverse debugging, from our own practical use of one such tool, Wind River Simics.

Index Terms— Software Debugging, Computer Simulation, Review, Computing History

1   INTRODUCTION

In this paper, actual *reverse debugging* is defined by the ability of a debugger to plant a breakpoint in a program and then proceed backwards in time until the breakpoint triggers. Reverse debugging is related to various implementation techniques such as record-replay, tracing, deterministic reexecution, reverse execution, checkpointing (or shapshotting), and program scheduling control, but it is really an application of those techniques to solve a debug problem that is core to the issue. Reverse debugging is sometimes known as *bidirectional debugging*.

As shown in Figure 1, reverse debugging is quite different from classic *cyclic debugging*, in which you run and rerun a failing program under debugger control to diagnose a failure. For cyclic debugging to be practical, it is pretty much required that each run of the programs behaves and fails in the same way.

For cyclic debugging to work, program under investigation has to be fundamentally deterministic. This is usually the case for non-interactive single-threaded programs, but not the case for real-time programs, parallel programs, or programs that involve some kind of asynchronous input/output (including files that change their contents between runs). In such circumstances, reproducing an error by rerunning a program is likely to fail (i.e., not hit the bug), or even hit different errors than the initial error

that prompted the debug session. Attaching a debugger often disturbs the timing of a parallel program, easily masking errors (so called *Heisenbugs*, where the act of observation changes the system to hide the bug).

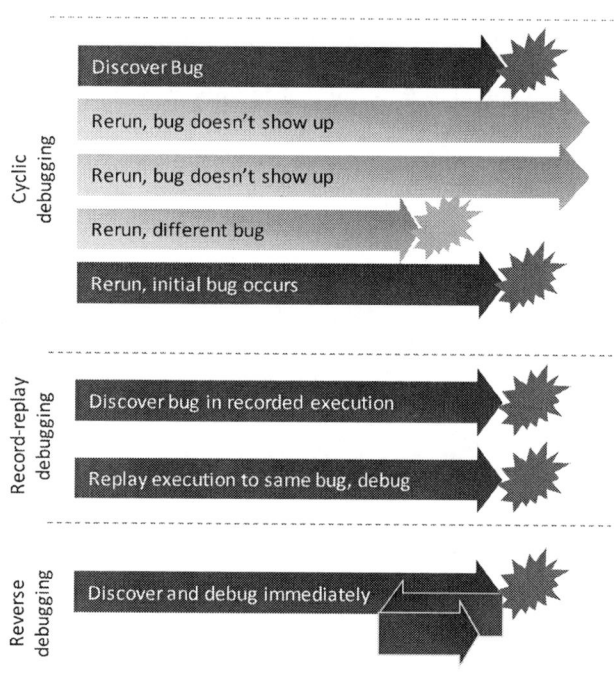

Figure 1 Fundamental debug techniques

Reverse debugging is one approach to tackle the debug of intermittent bugs, based on the idea of rather than trying to reproduce the issue in a separate run from the beginning you work inside the run that has already failed, and reverse back into its execution to diagnose the issue. This does require the ability to reproduce the past, and how this can be achieved is the topic of this paper.

An intermediate form between cyclic and reverse debugging is *record-replay debugging*. In record-replay, a non-deterministic execution is recorded and later replayed. Debugging is still performed in the forward direction: you cannot go back in time to a previous point in the program execution without restarting the replay run, nor can you trigger breakpoints backwards in time. Record-replay debug allows cyclic debugging to be applied to non-deterministic (but still recordable) programs. Record-replay is easier to implement, since it tends to require fewer changes to the debugger core than actual reverse debugging.

1.1 Going to a Past State

Fundamentally, it is impossible to actually reverse the execution of a computer program. Many machine instructions destroy information; just consider an operation such as XORing a register with itself or writing a new value to a memory location. There is no way to take the state after the operation and infer the state before the operation. Thus, we need to reconstruct history from saved information to obtain the past state of a program or computer system.

There are two ways to reconstruct past state. Either, we *record* it completely in a log, or we *reconstruct* it by executing forward from some saved state. Reconstruction only needs to record the information that cannot be reconstructed, which generates far smaller logs. Most practical solutions have chosen the reconstruction approach as the method to obtain the details of the system state at some particular point in time. Complete log recordings (traces) are feasible for some embedded systems, where extensive trace capabilities exist in hardware, but tend to be too slow if implemented in software.

1.2 Debugger Scope

The scope of reverse debugger solutions varies depending on the problems that the debugger wants to solve and the nature of the implementation. Figure 2 shows an overview of the possible scopes of a debugger, as to what is included in the reverse debug process (i.e., what is reversed in the sense that its past state is presented by the debugger).

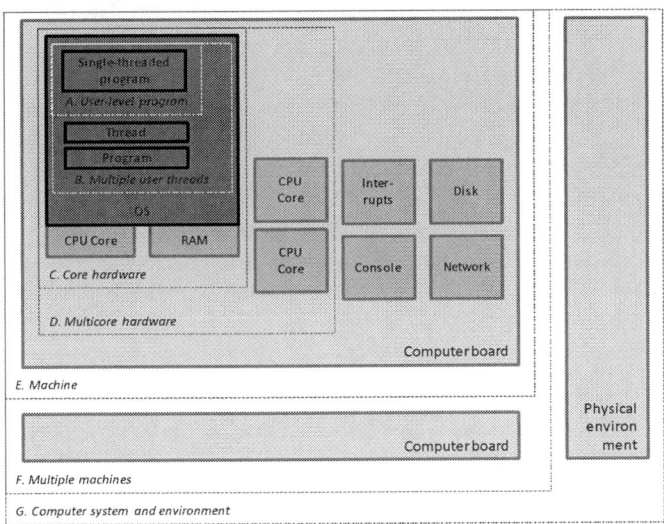

Figure 2 Possible Debugger Scope

User-level reverse debugging works on a single application (*A* and *B* in Figure 2). User-level debug can often be achieved in simpler ways than system-level debug,

but suffers from some limitations and can get very complex when dealing with IO.

System-level reverse debugging works not just on a user-level program, but rather on the scope of a full machine (minimally the operating system). Scopes C to G in Figure 2 are all system-level scopes.

An additional issue is the handling of *multiple* software threads or *multiple* hardware processors. It is in general much simpler to debug and reverse a single thread of execution than multiple threads. However, handling multiple threads is pretty much necessary today.

A related issue is the question of the code being run. Do we run *instrumented or unchanged programs* in the debugger? Instrumentation can help collect the information needed to construct the past state of a program. Example instrumentation approaches are compiling programs with extra instrumentation, using instrumented OS libraries, or special scheduling modes. Typically, running the same program binaries as on the real system is better.

Cross-target or host-based. Does the debugger run on the same machine as the program being debugged, or does it target a remote (or simulated) system? Normally for embedded systems, cross-target debug is the most interesting, while for desktop and server systems, host-based is the most natural.

1.3 History and Review

The idea of reverse debugging has been with us since the very beginning of computer programming, but was long dismissed as basically impossible or at least totally impractical. Advances in memory size, computing power, and debugging technology have made it feasible in recent years, even for fairly large systems and software sets.

The first presentation of a practical reverse debugging system that included the ability to step backwards and run backwards to breakpoints appeared in 2000 [1]. Before this, all prior research had focused on how to save and reconstruct old program state for record-replay debugging.

This paper will review the technology issues associated with reverse debugging, record-replay debugging, and related techniques. The presentation is structured after the technological issues, rather than presenting work in chronological order.

2 RECORD-REPLAY SOLUTIONS

There are a number of solutions that implement record-replay debugging, but not actual backwards step and/or breakpoints and thus do not quite qualify as reverse debuggers.

The simulator used in the *Data General Eagle* project in 1980 featured the ability to record and replay microcode executions in order to facilitate debugging [2]. It seems that this kind of trace and replay is a common feature of many

29

older simulation systems, as the idea is fairly obvious and the storage requirements for small-scale simulations modest.

The *Instant Replay* debugger from 1987 allowed the deterministic replay of parallel user-level programs on the BBN Butterfly parallel computer [3]. It required that programs adhered to certain communications styles to work, and relied on instrumenting the OS to capture all thread interactions. By replaying the thread interactions, *reconstruction* could be used to recreate the actual variable values of each thread. This was still considered a useful debugger for issues related threaded programs, in particular deterministic threaded bugs.

The *repeatable scheduling* system from 1996 instrumented system libraries and applied binary code modification to enable deterministic replay of multithreaded applications on a single processor [4]. Quite similar to Instant Replay, but with no requirements on how programs are coded, and no special debugger built around it.

Microsoft Visual Studio 2010 IntelliTrace records a history while debugging programs running on top of the .net CLR virtual machine [5]. It only collects selected data and lets the user move backwards in time to selected points to view past state. A nice feature is the ability to send the history file to other developers to share a debug session.

A weaker form of record-replay debugging is to replay only the stimuli known to cause an issue. In particular in networked systems, it is common practice to subject systems to a generated packet stream for testing (including fuzz testing), and then to replay the same packet stream in case the system fails. Packet streams or other input streams can also be captured on live deployed systems, and brought back to the developer labs to attempt to recreate issues from the field. Such trace replay does not really replay the same *system execution*, but it often works to recreate bugs that depend (mostly) on the nature of the packet stream.

3 TRACE-BASED REVERSE DEBUGGING

Trace-based debug solutions are a special case in that they simply record "everything". Most trace-based approaches use hardware debug support to capture a trace of the execution of a system, at level *C* in Figure 2. Then, in a separate step, a debugger reads the trace and performs reverse debug operations on the trace.

The scope of a trace-based solution is limited by what can be traced in a coherent way. Thus far, this has practically limited it to single processors, as time-stamped coherent traces from multiple processors are very difficult to obtain (that would be level *D* in Figure 2). As debug hardware improves, it is possible that this can be extended to an entire SoC including IO, which would be similar to *E* in Figure 2. The reverse time window is also limited by the capacity of the trace hardware to accept complete trace data.

Hardware trace-based approaches are cross-target, partially system-level (basically, only the processor),

multiple-thread, and can work on unchanged programs. The debugger also needs to recreate the target OS state from the trace in order to apply debug operations to OS abstractions.

To implement reverse breakpoints, the debugger scans backwards over the trace until some condition is satisfied. The implementation of reverse is all in the debugger core and its user interface, it does not affect the debugger backend.

The first practical commercial reverse debugger, the *Green Hills Time Machine* [6], was based on trace-based reverse, and came to market in 2003. *Lauterbach CTS* is another hardware-based debugger with reverse abilities [7], but apparently it does not feature reverse *breakpoints* and therefore is probably better classified as a record-replay debugger rather than a true reverse debug system.

There are also a few complete-trace systems based on software, which operate at level *A* or *B* in Figure 2.

The *Omniscient Debugger* presented in 2003 targeted programs running on top of a Java Virtual Machine [8]. It instruments the target program to log all state changes, and used a big lock to serialize execution of multiple threads. The implementation is admitted to be almost worst-case inefficient, but the debugger was still considered useful for some real work. The debug UI did feature backwards breakpoints, making it a true reverse debugger. This approach was user-level, instrumented programs, multiple threads, and host-based.

The *gdb 7.0* built-in record target which appeared in 2009 is a trace-based reverse debugging solution. It records the effect of every target instruction in a single target thread. It is very slow, but provides the only open-source and free solution for reverse debugging. It lacks a concept of time, and a user cannot move the execution to a particular point in time or run backwards for a certain time. The availability of gdb was also crucial to getting reverse execution features into mainstream Eclipse, in 2009.

Using the gdb-MI or gdb-serial remote protocols, gdb can also be used as a frontend to other commercial reversible debugger backends such as *Simics* [11] and *UndoDB* [12]. In this case, the gdb debugger does not know how the backend implements reverse, only that it understands the commands required to perform reverse. Thus, gdb offers a general frontend that can be used with multiple backends.

4 RECONSTRUCTION-BASED REVERSE DEBUGGING

Figure 3 shows the basic technique used to reach a certain previous point in time in a system execution in a reconstruction-based reverse debugger. The approach works by first setting the system state to a starting point that can be reliable reproduced, and which is *before* the desired target time. Such a system restoration capability is known as *checkpointing* or *snapshotting*, and we will call it checkpointing in this paper. The system then executes forward until the desired point in time has been reached. In the simplest case, a single checkpoint is maintained at the start of execution [9]. To reduce the user's waiting time most reverse execution approaches add additional checkpoints during the execution forward.

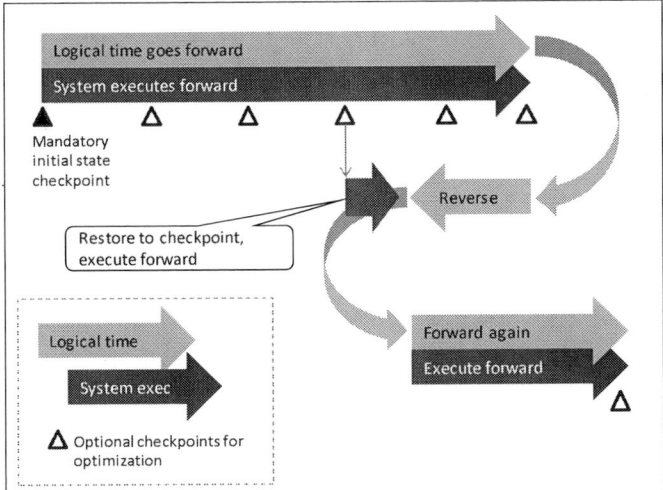

Figure 3 Reconstruction-based Reverse Principle

The execution forward has to be *deterministic* so that we always reach the same final state. If the system is subject to any asynchronous inputs, these have to be provided during the forward execution in order to make it possible to reach the same final state. Such inputs have to be provided with total precision in both data contents and timing, or the execution will diverge and become useless (at least for the purpose of reverse debugging).

4.1 Managing Time

To function, a reverse debugger needs to have a concept of time. Time is used to stop the reexecution process at the right point in time, and to trigger recorded inputs. How time is represented depends on the recording and reconstruction technology used. The most common time bases are executed instruction counts and clock cycles.

In systems with multiple threads or processors, time has to be ordered between the concurrent processes. How to implement the ordering intimately depends on the implementation of the reverse debugging system itself. For example, partial orders based on thread interaction [13], a total serialization of all threads using a single lock [8], or the scheduling scheme of a virtual platform [11] have all been used.

Most (but not all) reverse debug systems exposes time to the user. Typical operations available include being able to mark certain points in the execution to later come back to them (bookmarking), and running backwards or forwards for a certain amount of time. Some debuggers maintain a stack of past stopping points, allowing a user to logically browse backwards in their examination of the debuggee.

4.2 Reverse Step and Breakpoints

The two main operations you want in a reverse debugger is to be able to step backwards and run backwards until a breakpoint hits. The fundamental methods for this were presented as part of the *Bidirectional Debugger* in 2000 [1].

To step backwards, you essentially use the reconstruction approach to run one step short of the current position. You note the current time t in the system, and ask the reconstruction system to get to time $t - 1$ (in whatever time units are being used). For systems with multiple threads of control, the step can either be global or within the current thread, that depends on the debugger user interface decisions.

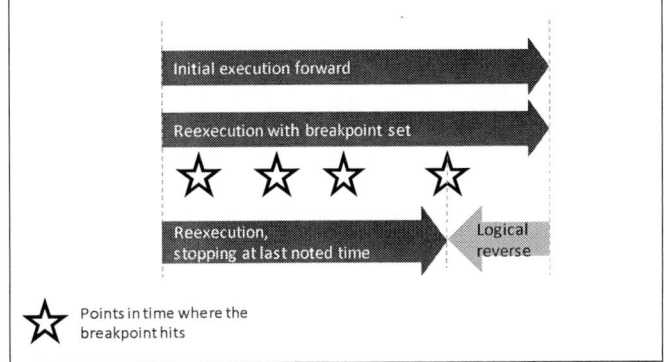

Figure 4 Reconstruction Reverse Breakpoint

To go backwards until a breakpoint hits, a somewhat more complicated approach is needed. All published work use the technique shown in Figure 4. The state is rewound to a checkpoint and then the system is executed forward once with breakpoints set. When a breakpoint hits during this rerun, the time is noted, and execution is resumed without notifying the user. Once the debugger reaches the current point in time, it stops the execution. It then goes back to the start again, and executes forward to the time of the last breakpoint hit seen during the rerun. Thus, it is enough to rerun the program twice to go back to the previous breakpoint. This is far better than stepping backwards a step at a time and checkpoint for breakpoint conditions, which would be quadratic in complexity and thus unusably slow. When using multiple checkpoints spread out in time, the strategy is normally to work backwards one checkpoint at a time.

4.3 Reversible State

The question about just which state to consider part of the reversing process is fundamental to the implementation of a reverse debugger based on reconstruction. Reconstruction can be applied to any level in Figure 2.

Obviously, for any computer program, the core state is the processor registers and memory system used to store the variables, call stack, current program location, and other data that the program operates on. From a hardware perspective, this is C or D in Figure 2.

When programs interact with the external world, things get more complicated, however. Fundamentally, as noted above, the program that we are reversing needs to get the same inputs at the same points in time that it got in its initial run. This has to be based on recording anything that cannot be reconstructed. How this is implemented varies widely between different reverse debugging systems, and is arguably the point where the most diversity is seen. If we look at Figure 2, we basically have to record any information that crosses the boundary between what we are able to reverse and the outside.

Thread interactions between threads in an OS are an important part of user-level reverse debugging systems. Both data exchange and synchronization have to be recorded. It is also necessary to record the asynchronous task switches where the OS interrupts a thread to schedule another thread, since the order of access to shared data is very important to reproduce to exactly reproduce a parallel program execution. For a *system-level* solution, thread interactions are implicitly reproduced by the reproduction of the operating system execution on top of the hardware.

An interesting twist on the management of thread interaction replay is to bend the rules a bit and change the order of replay of certain segments of code. This can be used to analyze the behavior of program by comparing the results from different replay orderings [14].

Console input and output is most commonly handled by replaying input data and ignoring output data during the reconstruction phase. This means that the display of a program (text or graphics) gets frozen at its final state [12]. This is safe but a bit unsatisfying in cases where output is of interest to the user. To reverse the contents of a display, the reverse debugger has to have control over the output system, which is the case for tools like virtual machines [9] and virtual platforms [11], but rarely for user-level soluions.

*File system*s. For a user-level solution, files leave a record of a program execution that is permanent from run to run of a program. Thus, file access has to be in some way virtualized or redirected so that a program always sees the same version of the file during its reconstruction phase. In a system-level reverse debugger, it is simply a matter of including the disk as part of the system state that needs to be reversed, making it no different from memory [9][11].

Network traffic, *sensor inputs*, *interrupts* from external sources, and similar interactions that reach out of level E in Figure 2 are most commonly handled by recording all inputs. Outputs are typically suppressed during reexecution, as the outside world does not expect or know how to deal with the information anyway. In reverse debugging systems based on full-system simulation and virtual platforms, it is possible to include the entire system including models of the environment into a single reversible whole [11].

4.4 Gaining Control

A key problem that needs to be solved for reconstruction-based reverse debugging is how to get control over the target system in order to be able to coerce it to repeat its execution. Standard computer systems are not repeatable or controllable to any useful extent, and thus some kind of intermediate layer has to be introduced to grab control.

Instrumenting the standard system libraries to capture inputs, outputs, and thread interactions is a necessary part of any user-level recording approach [1][3][4][12][13]. To handle multithreaded code, help from the OS is usually needed to capture events such as thread scheduling and descheduling [3][13]. Usually, it is also necessary to change the user code itself, to add time counters, record values of stores and loads, or coerce the reexecution of a program to do the right thing [1][4][13].

Another approach is to work on programs that are running on top of a language virtual machine (such as a JVM) [5][8]. A VM offers great control over the target software and its interface to the world, as everything that is done is fundamentally interpreted by the VM.

We can also use virtual platforms or virtual machines, and run the target software stack including the operating system. This has been used with both run-time virtual machines like UML [9] and VmWare [17], and development-oriented full-system virtual platforms like Simics [11]. A virtual platform reexecutes at the hardware level, and the software including the operating is reversed as a result. It is in many ways the cleanest solution, but also one that is hard to implement and get to perform well.

4.5 Implementations

Reconstruction of state seems to have been first implemented in the *Instant Replay* system [3], but that was really a record-replay rather than true reverse debug system.

The *Bidirectional Debugger* presented 2000 [1] used Unix fork calls to create checkpoints, and instrumented system libraries and program code to facilitate reexecution. Single-threaded and user-level.

ReVirt from 2002 used user-mode Linux (UML) paravirtual machines to implement a reverse debugger for operating systems [9][10]. The OS needs to be modified to run on UML, but not to run with the debugger. The paper [10] nicely presents a number of examples for when reverse debugging really helps. The solution is level E in Figure 2, but only for a single processor.

Simics Hindsight from 2005 used a full-system virtual platform to implement reverse debugging [11]. System-level, multiprocessor and multi-machine, unmodified software stack. A unique aspect of Simics is that it can reverse a system out to level G in Figure 2, and it was the first commercial multiprocessor reverse debugger.

UndoDB from 2006 was the first commercial user-level reverse debugger [12]. It runs on Linux, and can handle multiple threads (apparently by serializing their execution). It uses a novel time base, simulated nanoseconds, which increases by at least one between instructions, but does not correspond to any real time.

The Microsoft *iDNA* framework and the time travel debugger built on top of it was presented in 2006 [13][14]. It is a user-level, multiple thread solution that uses a combination of emulation and binary instrumentation to capture the data needed for replaying a program in a post-mortem debugger.

A small subset of the *Qemu* emulator was used to demonstrate reverse debugging in 2007 [15]. This work targeted the semihosting variant of Qemu, and thus was really user-level (not system-level), single-processor. Based on gdb, this debugger introduced the "change direction" user interface found in *gdb 7.0 reverse debugging*. All other reverse debugging solutions use the more natural and direct reverse (or backwards) step and run commands.

The RogueWave *TotalView* debugger added support for reverse debugging of multithreaded user-level programs in 2008, with *ReplayEngine* [16]. The debugger allows a user to step back in time, as well as run backwards to a certain line (which is a backwards execution breakpoint) and jump to a certain point in time. It requires heavy instrumentation of the runtime system used, and supports a few common HPC communications APIs such an OpenMPI.

VmWare Workstation added reverse debugging in 2008 [17]. System-level, single-processor solution that supported data breakpoints in revers. Based on recording the non-deterministic aspects of a single-processor VmWare machine, with limitations to which kinds of devices could be used. It noteworthy that VmWare dropped this support with VmWare Workstation 8 in 2011, presumably since the requirements of reverse execution interfered with the requirements for normal VM use cases.

5   FINAL REMARKS

This paper has offered a review of the current state and historical growth of reverse debugging technology. Over time, many different implementations have been tried, and some have made it all the way to users and the marketplace. The implementations differ in their scope, target systems, and capabilities, but all offer something better than traditional cyclic debugging. The space of this paper has only allowed a broad overview of the field, and the reader is encouraged to visit the referenced sources for more information.

6   REFERENCES

[1] Boothe, B, "Efficient Algorithms for Bidirectional Debugging", *Proc. ACM SIGPLAN 2000 Conference on Programming Language Design and Implementation (PLDI)*, pp. 299-310, May 2000.

[2] Kidder, T, *The Soul of a New Machine*, Hachette Book Group USA, NY, 1981.

[3] LeBlanc, T, and Mellor-Crummey, J, "Debugging Parallel Programs with Instant Replay", *IEEE Transactions on Computers*, pp. 471-482, Volume 36, Issue 4, April 1987

[4] Russinovich, M, and Cogswell, B, "Replay for concurrent non-deterministic shared-memory applications", *Proc. ACM SIGPLAN 1996 Conference on Programming Language Design and Implementation (PLDI)*, pp. 258-266, June 1996.

[5] Huff, Ian, "IntelliTrace in Visual Studion 2010 Ultimate", *MSDN Blogs*, May 13, 2009
(http://blogs.msdn.com/b/ianhu/archive/2009/05/13/historical-debugging-in-visual-studio-team-system-2010.aspx)

[6] Lindahl, M, "The Device Software Engineer's Best Friend", *IEEE Computer*, May 2006.

[7] Lauterbach Context Tracking Systems, http://www.lauterbach.com/cts.html, 2005.

[8] Lewis, B, and Ducasse, M, "Using Events to Debug Java Programs Backwards in Time", *Proc. of the ACM SIGPLAN 2003 Conference on Object-oriented programming, systems, languages, and applications (OOPSLA)*, pp. 96-97, October 2003.

[9] Dunlap, G, King, S, Cinar, S, Bazrai, M, and Chen, P, ""ReVirt: enabling intrusion analysis through virtual-machine logging and replay", *Proc. of the 5th symposium on Operating systems design and implementation (OSDI)*, pp. 211-224, 2002.

[10] King, S, Dunlap, G, and Chen, P, "Debugging Operating Systems with Time-Traveling Virtual Machines", *Proceedings of USENIX 2005 Annual Technical Conference*, pp. 1-15, 2005.

[11] Engblom, J, Aarno, D, and Werner, B, "Full-System Simulation from Embedded to High-Performance Systems", in *Processor and System-on-Chip Simulation*, Leupers, Rainer and Temam, Olivier (eds), pp. 25-45, Springer Verlag, 2010.

[12] UndoDB Man Page, http://undo-software.com/product/undodb-man-page, visited July 2012.

[13] Bhansali, S, et. al, "Framework for Instruction-level Tracing and Analysis of Program Executions", *Proc. of the 2nd International Conference on Virtual Execution Environments (VEE)*, ACM Press, June 2006.

[14] Narayanasamy, S, et. al, "Automatically Classifying Benign and Harmful Data Races Using Replay Analysis", *Proc. ACM SIGPLAN 2007 Conference on Programming Language Design and Implementation (PLDI)*, June 2007.

[15] Jacobowitz, D, and Brook, P, "Reversible Debugging", *GCC Developer's Summit* 2007.

[16] Gottbrath, C, "Reverse Debugging with the TotalView Debugger", *Cray User Group Conference*, Helsinki, Finland, May 2009.

[17] Lewis, E, "VMware Workstation 6.5: Reverse and Replay Debugging is Here!"
(http://www.replaydebugging.com/2008/08/vmware-workstation-65-reverse-and.html), August 2008

Application of Timing Variation Modeling to Speedpath Diagnosis

Mehdi Dehbashi*

*Institute of Computer Science, University of Bremen
28359 Bremen, Germany
Email: dehbashi@informatik.uni-bremen.de

Görschwin Fey*†

†Institute of Space Systems, German Aerospace Center
28359 Bremen, Germany
Email: goerschwin.fey@dlr.de

Abstract—The impact of timing variations on the performance of *Very-Large-Scale Integrated* (VLSI) circuits is increasing as the feature sizes shrink down into the nanometer scale. Timing variations induced by process, environmental or other effects may lead to a failing speedpath. In this paper, first a functional model of circuit timing is constituted. Then, timing variations are added to the model. Afterwards, this model is utilized to diagnose failing speedpaths.

Keywords—diagnosis, speedpath, timing variation

I. INTRODUCTION

Diagnosis of failing speedpaths is a major challenge in developing VLSI circuits as timing variations induced by process and environmental effects increase. A path which limits the frequency of a circuit is called a *speedpath* [1] [2]. A speedpath that fails timing constraints at the post-silicon stage is called *failing speedpath* [3].

The correct behavior of a circuit is verified at the stage of post-silicon validation by applying test vectors to the chip. when an error is detected, the debugging starts to diagnose root causes of the error. But this process is often a manual task which consumes a significant portion of the IC development cycle. Therefore, automated debugging approaches are necessary to speed up this process.

In the recent years, due to statistical variations imposed by process variations, *Statistical Static Timing Analysis* (SSTA) methods have been proposed [4]. SSTA methods analyze timing behavior of a circuit under statistical variations. A formal procedure based on integer linear programming is presented in [3] to diagnose segments of failing speedpaths. A variational model obtained from parameterized static timing analysis is used in [5] to create a cost function. Then, a branch and bound approach using this cost function determines probable failing speedpaths.

An approach based on trace buffers is presented in [6] to debug failing speedpaths. Trace buffers are used as a hardware structure to provide real-time observability to speedpaths during the normal operation of a chip. A scan-based hardware structure is used in [7] to debug failing speedpaths. The approach explores debug techniques based on at-speed

This work has been funded in part by the German Research Foundation (DFG, grant no. FE 797/6-1).

scan test patterns. Failing traces are analyzed at slower-than-nominal clock frequencies in [8] to enhance the diagnosis resolution. In [9], failing speedpaths are isolated by applying clock shrinking on a tester and using a CAD methodology. The work in [10] models the timing behavior of a circuit and timing variations in a functional domain. The model is used to analyze the functional behavior of a circuit under timing variations.

In this paper, we utilize the timing variation model of [10] to diagnose failing speedpaths. First, the timing behavior of a circuit is converted into a functional domain based on a discrete model of time units. Modeling the timing behavior of a circuit in the functional domain allows the formal methods to comprehensively analyze the timing effects of a circuit. Timing variations are also modeled varying the value of a signal by an accuracy of one time unit. Given an erroneous trace obtained from a testbench, the created model is constrained to the erroneous trace in order to diagnose failing speedpaths. Our approach uses a SAT solver as an underlying engine.

The remainder of this paper is organized as follows. Our diagnosis approach is presented in Section II. Section III presents experimental results on benchmark circuits. The last section concludes the work.

II. APPROACH

At the post-silicon stage, test vectors are applied to the chip and the clock period is reduced to detect the erroneous effects of timing variations. This step is called clock shrinking. The erroneous behavior of timing variations is observed on registers or outputs. An error is detected by comparing the output values of the chip with the nominal output values obtained from simulation at the specified clock period. The test vectors activating timing variation and the erroneous output constitute an *Erroneous Trace* (ET).

Given an erroneous trace obtained from the testbench due to timing variations, our goal is to diagnose failing speedpaths, i.e., to determine which speedpaths have failed due to a timing variation. In the approach, a chip is validated in a testbench using clock shrinking and test vectors. The output of the testbench is an erroneous trace.

To debug a circuit, first a *Time Accurate Model* (TAM) of the circuit is constructed according to a time unit. The TAM is

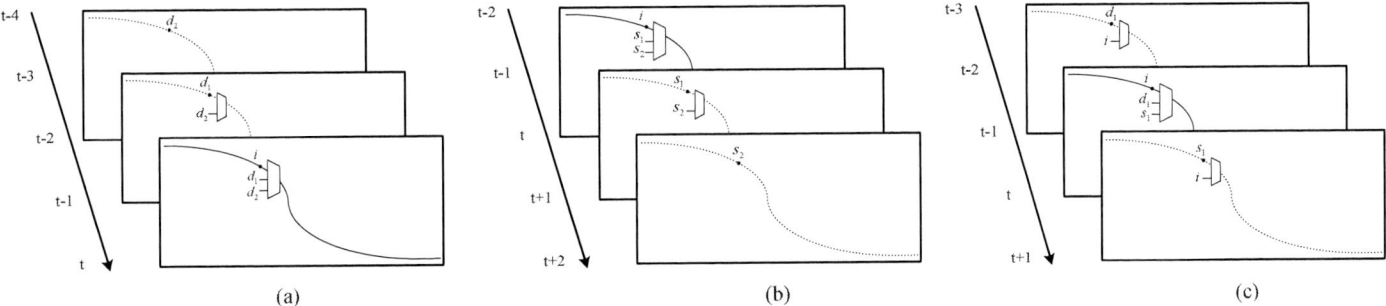

Fig. 1. (a) Slowdown (b) Speedup (c) Slowdown and Speedup

used in [10] to analyze the functional behavior of logic circuits under timing variations. The time unit specifies the diagnosis accuracy of the approach. To construct the TAM, first all gates are converted to *untimed gates*. Each untimed gate has a delay of one time unit. Buffers are inserted at the output of each original gate in order to convert an original gate to an untimed gate. After this step, all components in the circuit have a delay of one time unit. The new circuit is called *untimed circuit*. A TAM circuit is constructed from an untimed circuit. The underlying idea to construct a TAM is to use copies of a gate to represent the value of a gate at different points in time. Therefore, if an untimed gate is exercised several times at different time steps (e.g. due to reconvergent fanout), one copy of the untimed gate in each related time step is created. Having an erroneous trace and a TAM circuit, the diagnosis process starts to find failing speedpaths.

In this case, timing variations are added into the model. Timing variation may increase the component delay called slowdown, or may decrease the component delay called speedup. To model maximum timing variation V, V additional copies of the TAM are created. Figure 1 shows modeling timing variations when $V = 2$. Then, multiplexers are added to the TAM gate outputs to model timing variations by selecting the signal values from different time steps. Figure 1(a) shows modeling of slowdown. Speedup modeling is shown in Figure 1(b). Figure 1(c) is used for both slowdown and speedup. In Figure 1, values d_1 and d_2 denote a slowdown of one time unit and two time units for value of signal i. Values s_1 and s_2 denote a speedup of one time unit and two time units for value of signal i. The select lines of multiplexers are controlled by a constraint v. First v is 1. In this case, the inputs and outputs of the TAM is constrained to the erroneous trace ET and the diagnosis answers the following question: If there is a timing variation of one time unit, can the erroneous behavior of the corresponding ET be observed on the outputs? If there is no solution, the diagnosis function increases v. This process repeats until v reaches the maximum timing variation V.

The output of the diagnosis process is a set of fault candidates FCs. Each fault candidate includes the spatial and temporal information of a gate. All fault candidates together

show failing speedpaths. Then, these failing speedpaths are visualized on the schematic view of the circuit.

III. EXPERIMENTAL RESULTS

In this section, we evaluate our diagnosis approach empirically to debug the logic circuits under timing variations. The experiments are carried out on a Quad-Core AMD Phenom(tm) II X4 965 Processor (3.4 GHz, 8 GB main memory) running Linux. We use the combinational circuits of the ISCAS'85 benchmark suite to evaluate our approach. We synthesize our circuits using Synopsys Design Compiler with Nangate 45nm Open Cell Library [11].

The TAM-based diagnosis described in this paper is implemented using C++ in the WoLFram environment [12]. For the experiments, one time unit is $0.01\ ns$. MiniSAT is used as underlying SAT solver [13].

The simulation testbench is implemented using Verilog in ModelSim environment. Single slowdown faults of one time unit are injected in the circuit. Random test vectors are used in order to detect a slowdown in our testbench. Then, the erroneous trace is debugged by the approach.

Table I presents the experimental results. The table shows the circuit name (first column), the total number of gates (#Gates), the required run time in CPU seconds (Time), and the final number of fault candidates (#FC). Each fault candidate indicates that if a slowdown of one time unit at the appropriate time step on the output of the corresponding gate occurs, the erroneous behavior of the ET is created. The number of fault candidates (#FC) indicates the diagnosis accuracy of the diagnosis approach. A smaller number of #FC indicates a higher accuracy.

The second column in the table shows the total number of gates in the original circuit. By inserting buffers at the output of the original gates, it is converted to an untimed circuit. The total number of gates in the untimed circuits is shown in the third column. Afterwards, the TAM circuit is constructed. The fourth column shows the total number of gates in the TAM circuits. The required time to construct the TAM, to do diagnosis, and the total time are shown in columns five through seven.

TABLE I
TAM-BASED DEBUGGING

Circuit	#Gates when Time Unit = 0.01ns			Time (s)			#FC
	Original	Untimed	TAM	TAM	DBG	Total	
c17	6	26	35	0	194.97	194.97	2
c432	115	511	15446	135.01	1473.17	1608.18	20
c499	179	840	4358	4.6	212.58	217.18	2
c880	172	814	6483	14.95	1258.92	1273.87	17
c1355	238	1112	14338	93.33	2024.13	2117.46	26
c1908	142	658	5171	6.68	260.5	267.18	3
c2670	280	1296	8817	35.28	1391.03	1426.31	19
c3540	391	1792	50664	2347.64	1010.63	3358.27	10
c5315	632	3042	18283	290.24	1158.21	1448.45	16
c7552	772	3657	58468	3240.66	2093.9	5334.56	21

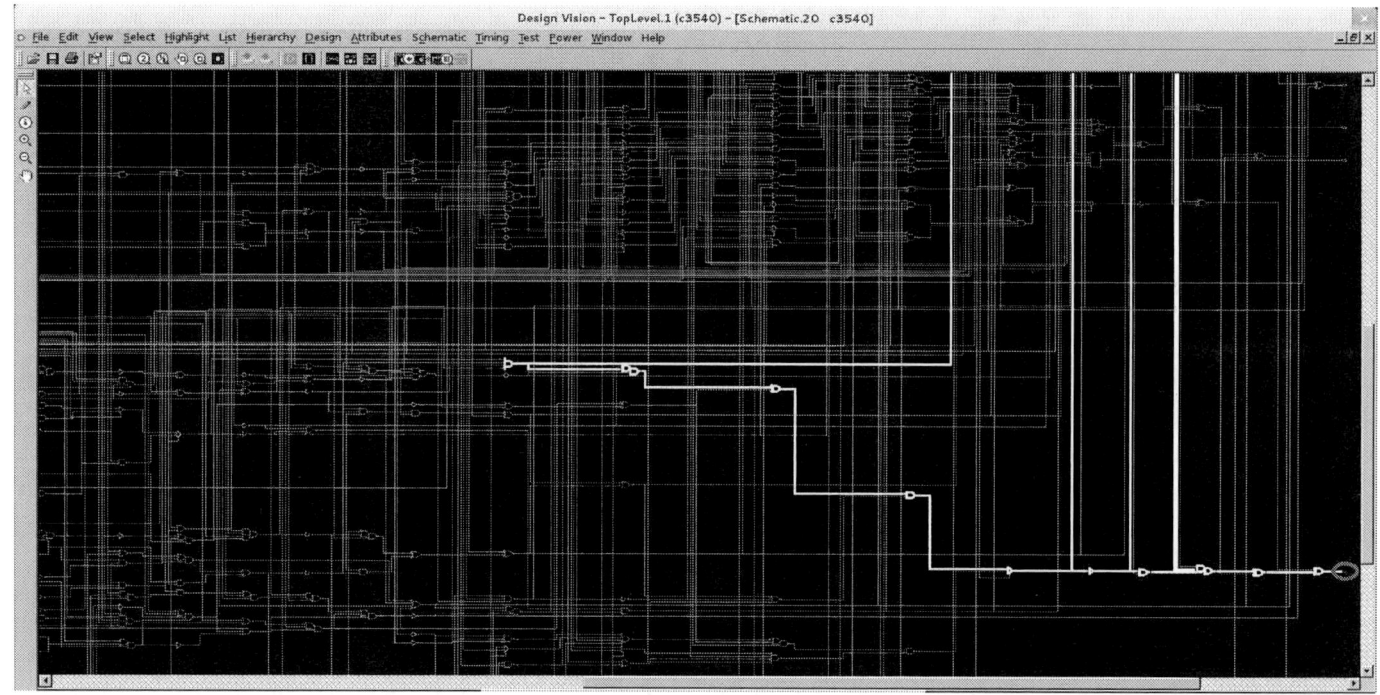

Fig. 2. Failing speedpath in circuit c3540

For circuit c432, the number of TAM gates is 15446, while the number of TAM gates is 4358 for circuit c499. This shows that in c432 there are more reconvergent fanouts in comparison to c499. The reconvergent fanouts increase the size of the TAM. The number of fault candidates for c432 is 20, while the number of fault candidates for c499 is 2.

In our experiments, all fault candidates together highlight failing speedpaths or some segments of failing speedpaths. Each fault candidate includes the location and the time of fault activation.

For circuit c3540, there are 10 fault candidates. They are visualized on the schematic view of the circuit in Figure 2 using *Synopsys Design Vision* [14]. The gates highlighted in white color show a segment of a speedpath which has violated the timing constraint. The red circle shows the output on which the error was observed. The approach was also applied to sequential circuit s298 from the ISCAS'89 benchmark suite. In this case, the erroneous behavior may be observed on internal registers of the circuit. Figure 3 shows the failing speedpath for the sequential circuit s298. The red circle shows a flipflop on which the error was observed. As the experiments show, our approach achieves a high diagnosis accuracy and can automatically extract failing speedpaths.

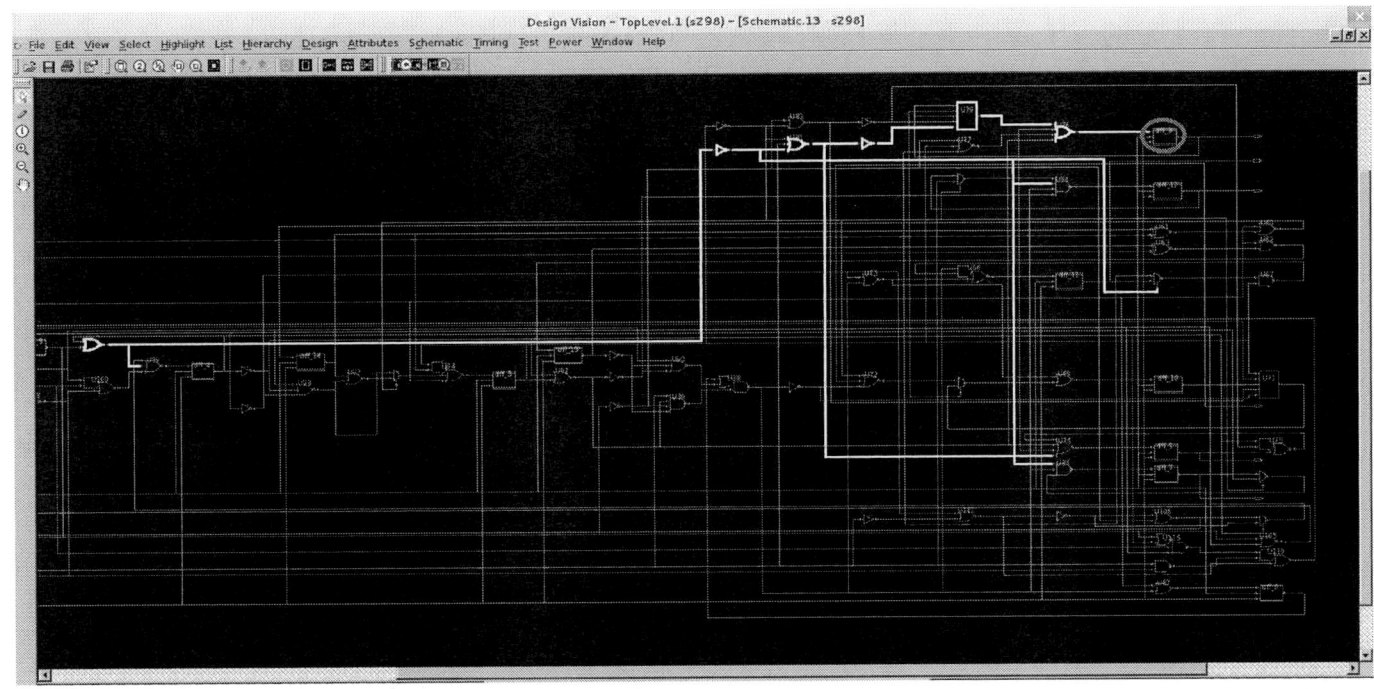

Fig. 3. Failing speedpath in circuit s298

IV. CONCLUSION

We presented an approach to diagnose speedpaths in logic circuits under timing variations. In the approach, first the timing behavior of a circuit is converted into the functional domain under a discrete model of time unit. Then, timing variation models are inserted into the functional domain. Afterwards, our diagnosis approach finds potential fault candidates including their spatial and temporal information.

REFERENCES

[1] P. Bastani, K. Killpack, L.-C. Wang, and E. Chiprout, "Speedpath prediction based on learning from a small set of examples," in *Design Automation Conf.*, 2008, pp. 217–222.

[2] L. Lee, L.-C. Wang, P. Parvathala, and T. M. Mak, "On silicon-based speed path identification," in *VLSI Test Symp.*, 2005, pp. 35–41.

[3] L. Xie, A. Davoodi, and K. K. Saluja, "Post-silicon diagnosis of segments of failing speedpaths due to manufacturing variations," in *Design Automation Conf.*, 2010, pp. 274–279.

[4] D. Blaauw, K. Chopra, A. Srivastava, and L. Scheffer, "Statistical timing analysis: From basic principles to state of the art," *IEEE Trans. on CAD of Integrated Circuits and Systems*, vol. 27, no. 4, pp. 589–607, 2008.

[5] S. Onaissi, K. R. Heloue, and F. N. Najm, "PSTA-based branch and bound approach to the silicon speedpath isolation problem," in *Int'l Conf. on CAD*, 2009, pp. 217–224.

[6] X. Liu and Q. Xu, "On signal tracing for debugging speedpath-related electrical errors in post-silicon validation," in *Asian Test Symposium*, 2010, pp. 243–248.

[7] J. Zeng, R. Guo, W.-T. Cheng, M. Mateja, J. Wang, K.-H. Tsai, and K. Amstutz. "Scan based speed-path debug for a microprocessor," in *European Test Symposium*, 2010, pp. 207–212.

[8] V. J. Mehta, M. Marek-Sadowska, K.-H. Tsai, and J. Rajski, "Timing-aware multiple-delay-fault diagnosis," *IEEE Trans. on CAD*, vol. 28, no. 2, pp. 245–258, 2009.

[9] K. Killpack, S. Natarajan, A. Krishnamachary, and P. Bastani, "Case study on speed failure causes in a microprocessor," *IEEE Design & Test of Computers*, vol. 25, no. 3, pp. 224–230, 2008.

[10] M. Dehbashi, G. Fey, K. Roy, and A. Raghunathan, "Functional analysis of circuits under timing variations," in *European Test Symposium*, 2012.

[11] *Nangate 45nm Open Cell Library*, http://www.nangate.com.

[12] A. Sülflow, U. Kühne, G. Fey, D. Große, and R. Drechsler, "WoLFram – a word level framework for formal verification," in *IEEE/IFIP Int'l Symposium on Rapid System Prototyping*, 2009, pp. 11–17.

[13] N. Eén and N. Sörensson, "An extensible SAT solver," in *SAT 2003*, ser. LNCS, vol. 2919, 2004, pp. 502–518.

[14] *Design Vision - Synopsys Inc.*

Vienna, Austria – September 19-20, 2012

Session 3: Verification and Virtual Prototyping

Co-Debug and Co-Verification Environment for Power Management System
Markus Winterholer (Cadence)

Scalable and Retargetable Debugger Architecture for Heterogeneous MPSoCs
Luis Gabriel Murillo, Julian Harnath, Rainer Leupers and Gerd Ascheid (RWTH Aachen)

Co-Simulation Framework for Variation Analysis of Radio Frequency Transceivers
Sumit Adhikari, Florian Schupfer and Christoph Grimm (Vienna University of Technology)

www.ecsi.org/s4d

Co-Debug and Co-Verification Environment for Power Management System

Markus Winterholer

Cadence GmbH, Feldkirchen, Germany
markus@cadence.com

ABSTRACT

Power management used to be a domain of mobile devices. Since energy bills and the environmental impact of greenhouse gas emissions from power plants are in everybody's mind, energy efficiency is a hot topic for each embedded system design. In the system design flow, trade off's between hardware and software functionality have a major impact on power saving, i.e. data compression, cryptography, highly utilized functions, etc. As a result, power management is distributed between hardware and software implementation features and therefore has to be developed and validated as early as possible in a single environment, requiring co-debug and co-verification capabilities. Additionally, power architectures add complexity to system debug and verification since each power mode has to be validated in combination with all valid features.

The presentation will introduce the complexity problem of a power management system distributed among hardware and software layers and will discuss the debug and verification challenges on different commonly used development platforms from virtual prototyping to the final embedded system.

Index Terms— co-debug, co-verification, embedded systems, power management

1. INTRODUCTION

Embedded systems should consider power management features in their design. Telecommunication products, electrical appliances, robots, medical devices and aircrafts are a few examples in this area. In these systems both hardware (HW) and software (SW) modules compose the power management system and, therefore, both should be well verified.

The main challenge in the area of low power verification is how to handle its complexity, since power management requires features from hardware, software and the integration of both together. The specification of the system clock requirements, controlling clock frequency and gating are critical in reducing the predefined factors of power consumption. On one hand, the embedded software can be extended to enable/disable and frequency switch clocks to various parts of the design as required. On the other hand, hardware control can be used to enable and disable clocks on detection of a functional request for a specific function. In both cases these are controlled through a central clock distribution circuit for clock gating, frequency mode switching, test bypass, etc. Additional clock phasing control can be introduced from this point to reduce peak power, and stress from high levels of simultaneous switching [1].

In order to verify the aforementioned functionality, new strategies and design techniques are required to address power management issues. These strategies need to address every level of abstraction within the design-cycle. The gains have a much higher magnitude when considered in the earlier phases of the design. For instance, at system and architectural levels a right decision can reach up to 10 or 20 times higher magnitude effects on power. On the other hand, in the later design phases (e.g., implementation or layout) only up to 20% of improvement might be achieved [1].

However, the most commonly used approaches to verify power management modules are focused in later design phases based on directed test approaches possibly taking advantage of co-debug and/or co-simulation solutions. This results in a high effort to create directed test vectors and critical corner case scenarios might go unnoticed. Furthermore, one of the main problems in power management verification is covering and stressing the integration of hardware and software modules. Random generation can automate test vector generation but is worthy only with constraints applied along with coverage measurement. Constraints are used to avoid the generation of illegal stimuli as well as to steer towards interesting scenarios and the coverage approach is used to measure the verification performance of the system.

Metric-driven verification (MDV) [2] (also known as coverage-driven verification [3]) has been used successfully in the hardware area, for instance in the e language [4]. Recently it has been extended to the embedded software [5-11]. Thus, this approach is a suitable approach to overcome the complexity that the power architecture adds to the verification process. Metric-driven verification is able to handle space growth with each power mode and to validate the device in all the modes. Additionally, MDV can generate test cases to cover all legal transitions and it ensures that modes are entered and exited without errors, which is one common cause of error in low power designs. Finally, MDV helps in providing documentation with the intent of the power architecture in order to ensure the communication across teams and functional groups, i.e., architect, system designer, RTL designer, and verification engineers.

In this paper we present a co-debug environment which is based on a metric-driven verification approach [12] in order to debug functionalities of hardware and software modules in a power management system.

2. DEBUG AND VERIFICATION ENVIRONMENT

A systematic way of debugging and verifying a system is metric-driven verification. The use of a verification plan and coverage metrics organize and manage the verification project, and optimize daily activities to reach verification closure. Executing regression suites produces a list of failing runs that typically represent bugs in the system to resolve, and coverage provides a measure of verification completion. Bugs are iteratively debugged fixed. Analyzing coverage holes provides insight into system scenarios that have not been generated, enabling the verification team to make adjustments to the verification environment to achieve more functional coverage [13].

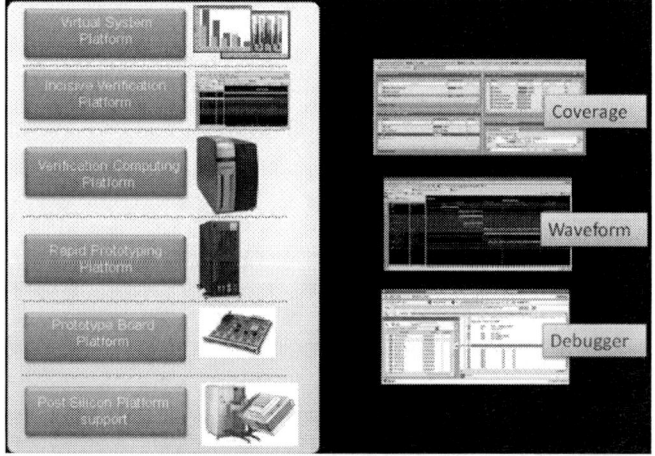

Figure 1: Platform connectivity

In order to enable the application of this methodology to corner case scenarios, which lie across the hardware software boundary, we need to have the same interface capabilities with the software running on an embedded processor as we do with the differing abstractions of hardware models. In particular we are interested in being

able to stimulate the embedded software, i.e. call its routines and drive its variables as well as monitoring its state. Figure 1 lists the various execution platforms used during the hardware and software design cycle. Ideally, failures have to be reproduced and debugged with a common user interface on all models.

The mechanism that is used to communicate with the software must be independent from the method of execution and the abstraction of the processor model. Here are some examples of the mechanisms that we typically see for modeling embedded processors [5].

- Host Code Execution (HCE): This technique uses the host workstation rather than a model of the actual processor to run the embedded software.
- Instruction Set Simulator (ISS): Most processor vendors provide an ISS model of the processor that provides a highly abstract model, which may be executed in the context of a hardware simulation.
- Register Transfer Level (RTL): One option is to have the real RTL for the processor in the simulation. In practice an RTL model is often replaced by an abstracted cycle accurate model that will provide accuracy with increased performance.
- Acceleration/Emulation: Often, for SoC level designs software based simulations are no longer suitable due to the scale of simulation required. In this case we may choose to accelerate or emulate the design including the processor.

For all of these cases we must ensure that the mechanism through which the verification environment communicates with the embedded software is consistent. For these reasons an independent approach, although more complex, is likely to be the most useful in real applications. This is the mechanism used in the Generic Software Adapter (GSA) and will be described further.

2.1 GENERIC SOFTWARE ADAPTER (GSA)

GSA [5-6] provides communication with the software running on the embedded processor and it does this in a model and processor independent manner. This is achieved through a mailbox (i.e. shared buffer) located in the processors memory map. This mailbox is written to and read from both by software running on the embedded processor and by the verification environment. The verification environment writes tasks to the mailbox so that an embedded software wrapper may notice these tasks and act upon them.

The mailbox gives the verification environment the ability to indirectly control and observe activity in the embedded

40

software. When the verification environment needs to make a call to a method in the embedded software, it firsts needs to check that the software is inactive and has been initialized. It will achieve this by reading the status from the mailbox. It will then set the parameters for the method that is to be called. These parameters may be of any type but are most commonly scalar types. Once the parameters have been set then a flag is set to indicate that the method should run. At this point the thread in the verification environment interfacing to the mailbox will suspend until the flag is set to indicate that the process has completed. Then the output parameters may be read and the call is completed.

By looking from the software side, the first step to be done is to initialize the mailbox. Once this has been completed the software may start monitoring the mailbox to see if any activities are being requested. During the monitoring process we are looking for method calls to be placed in the mailbox. The polling mechanism consists of a large case statement encompassing all of the methods that we might like to call. These will be defined manually or automatically. The call box indicates via an integer which method we want to execute and then the case statement makes the call to the relevant method, assigning parameters from the mailbox accordingly. When completed, the return value is written back to a mailbox and then completion is indicated via the activity mailbox flag. A similar process may be used to monitor software state variables and to provide a callback mechanism.

3. POWER MANAGEMENT VERIFICATION

The verification process of a power management system should follow iteratively four main phases of planning, execution, implementation and analysis.

3.1 PLANNING

Verification Planning is a methodology that defines how to measure the power modes, scenarios, and features per mode. Additionally, it documents how results of verification are measured considering, for instance, simulation coverage, directed tests, and formal analysis. It also provides a framework to reach consensus and define verification closure for this design. Figure 2 shows a typical design responsible for the power management. For this design we can define a verification plan consisting of:

- Power Control Module (PCM) validation: Focus on automation of assertions and coverage of the control signals and sequences. In particular handling of clocks and resets customized by the user.
- System level power metrics: Definition of system level sequences between modes; Checking expected

activities work in each mode and management of the complexity of multiple power modes.

- Domain/subsystem level: Assure that the enable and disable of power shutoff worked correctly; Checking the sequence beyond automatic monitors, including tests to make sure the restore worked correctly, or that specific transactions occurred after a power up. Additionally, ensure that standby mode worked as expected

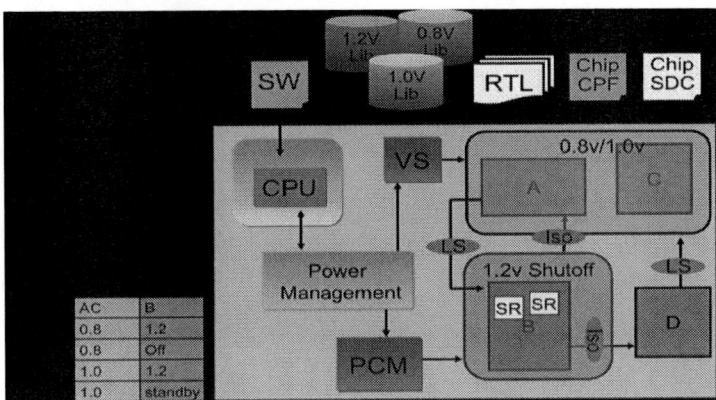

Figure 2: Power management module

3.2 INSTRUMENTATION

The instrumentation process consists of the development of an advanced power aware testbench, containing coverage metrics and assertions in order to verify power modes, behavior during each mode, power control logic and interface between domains.

Coverage metrics and assertions will help to ensure that each domain has executed all of its possible states and transitions. Example of states are High, Low, Standby, and Off. Examples of transitions are shown in Figure 3.

From	To
High	Low
High	Standby
High	Off
Low	High
Low	Standby
Standby	High
Standby	Low
Off	Low

Figure 3: Low power state transitions

3.3. EXECUTION

Formal or simulation-based approaches together with emulation approaches can be used.

3.4 MEASURE, REPORT, AND ANALYZE

Finally, the verification coverage has to be analyzed and the debugging process should be started in order to discover bugs.

4. POWER VERIFICATION SOLUTION WITH CADENCE'S INCISIVE VERIFICATION KIT

The Incisive Verification Kit is a complete platform designed for fast knowledge transfer of key verification methodologies. It contains HW and SW modules to develop a complete embedded system including power management features. It can be represented as a stack as shown in Figure 4 and detailed in the following subsections.

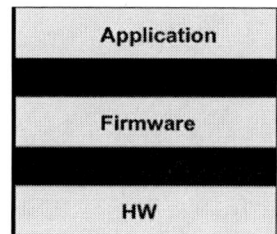

Figure 4: Power management stack

4.1 HARDWARE MANAGEMENT

The HW power management is able to support SW and key blocks that need power shutoff capabilities. As shown in Figure 5, Power Control Module can switch power shutoff domains and voltage-controlled oscillator (VCO) frequencies controlling some main modules such as: OR1K processor, instruction and data memories, Ethernet MACs, DMA and UART.

With the aforementioned reference design it is possible to verify 16 switchable power domains resulting in 37 system power modes, for instance:
- 4 MAC's switchable at 1.2V/off
- 1 Processor switchable at 1.2V/off
- 5 On chip Data memory switchable at 1.2V/Standby/off
- 1 On chip instruction memory switchable at 1.2V/off
- 1 DMA switchable at 1.2V/off
- 1 ALU switchable (hardware controlled) at 1.2V/off

- 1 UART switchable at 1.2V,1.0V, 0.8V, 0.6V and off
- 1 SMC switchable at 1.2V,1.0V, 0.8V, 0.6V and off
- 1 Default/core domain operating at 1.2V,1.0V, 0.8V and 0.6V
- 4 Switchable clock speeds

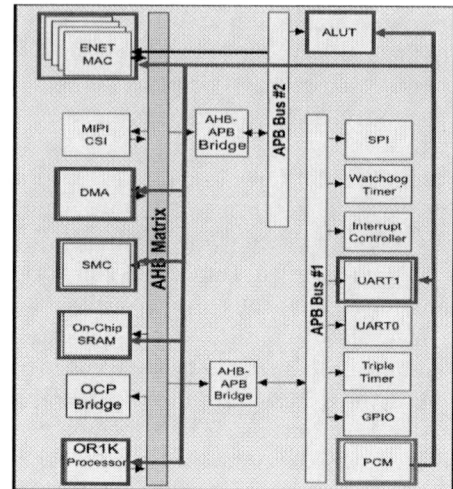

Figure 5: Incisive Verification Kit reference design

4.2 SOFTWARE MANAGEMENT

The SW power management is divided in two layers:
- **Firmware Layer** contains low-level knowledge regarding HW interface functionality and it automatically manages the power state without application intervention, for example in the following scenarios:
  • "Ethernet port has been inactive for a certain time, it will shutdown the power to that HW block".
  • "Traffic is detected on a powered-down port; Firmware will automatically power it up".
  • "2 out of 4 ports are powered down, by design this means we can slow down the CPU clock as well".
- **Application Layer** contains a finite state machine that is responsible for the power management. It is constituted of four states with the following functionality:
  • Configuration: It determines some static configurations, for instance, "Only ports 1-3 are enabled".
  • Operation: Operational mode is active whenever the system is fulfilling its designed function, for instance, "Forwarding ethernet packets".
  • Sleep: Sleep mode is when the processor is shutting down but memories and interfaces may be alive, for instance, "No ethernet traffic on any ports".

• Hibernation: Hibernation is where everything but the always-on power domain is active, for instance, "User has pressed the off button".

4.3 HW/SW CO-DEBUGGING

As aforementioned in Section 2, metric-driven verification has been recently extended to the embedded software. Generic software adapter enables a single verification environment for the co-debugging of hardware and software modules. Initially the system is executed without the CPU and software, which allows maximum control of bus activity and all types of transfers can be debugged. When CPU is reinserted, same sequences as before can be used but now drive software routines. This enables also the reuse of the verification environment and developed tests in order to reproduce the same scenario.

GSA needs to communicate with the software running on the embedded processor and it needs to do this in a model and processor independent manner. This is achieved through a mailbox located in the processors memory map. This mailbox is written to and read from both by software running on the embedded processor and by the verification environment, as we can see in Figure 6. The verification environment will write tasks to the mailbox so that an embedded software wrapper may notice these tasks and act upon them.

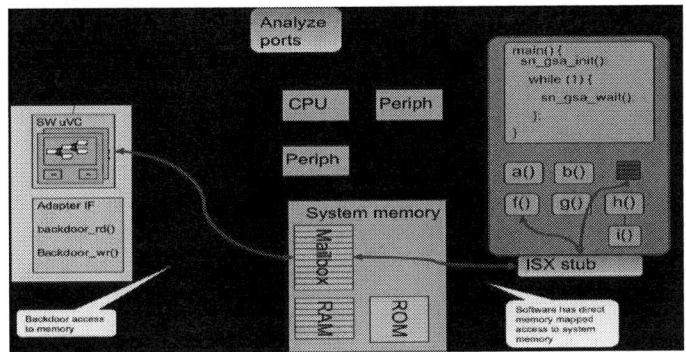

Figure 6: Hardware/software communication

4.4 MONITORING OF POWER AWARE VERIFICATION

The coverage model allows us to measure the verification progress in the power management design and also to dynamically adapt the sequence generation. The Incisive Verification Kit offers three coverage analyzes: 1) Item coverage shows if all legal values of power modes have been tested; 2) Transition coverage is applied for state machines and expresses which legal transitions among the power modes have been covered; 3) Cross coverage provides the cross product of basic items or transitions and shows if this

combination has occurred. This option is very useful to cover the interaction between HW and SW events.

In Figure 7 we can see the implementation of the power state coverage. We need to define a coverage group (i.e., power modes and mode transition) that contains a list of data items for which data is collected over the time. The data information is sampled based on events declared in the same unit. Besides, the sampling data can be split in ranges in order to control either which values has been tested or if coverage holes still exist.

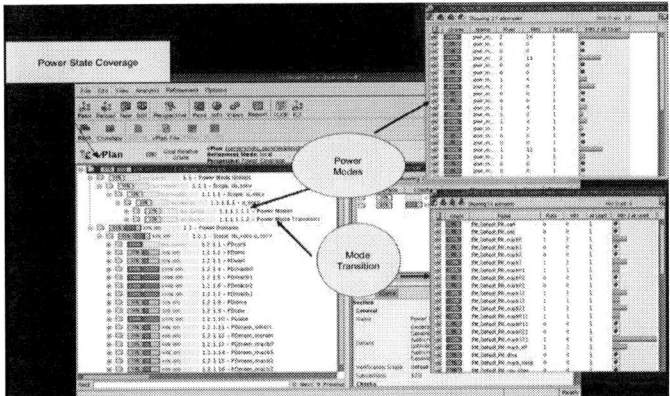

Figure 7: Coverage for power aware verification

5. CONCLUSION

In this work we presented a detailed novel approach for debugging and verification of power management using metric-driven verification technique. Based on this approach, we were able to verify and to cover difficult corner cases in both hardware and software modules. Coverage of power management is the main advantage of the solution with Cadence's Incisive Verification Kit, because one can avoid the coverage holes at earlier phase of design. GSA enables a single verification environment for co-debug and co-verification of hardware and software modules. This enables also the reuse of the verification environment and developed tests.

6. REFERENCES

[1] Cadence Design Systems, Inc. Low-power verification flow, URL: http://www.cadence.com/products/fv/pages/lp_flow.aspx, 2011.

[2] H. B. Carter and S. G. Hemmady, Metric Driven Design Verification: An Engineer's and Executive's Guide to First Pass Success, Springer, 2007.

[3] A. Piziali, Functional Verification Coverage Measurement and Analysis. Kluwer Academic Publishers, 2004.

[4] S. Iman and S. Joshi, The e Hardware Verification Language. Kluwer Academic Publishers, 2004.

[5] M. Winterholer, Transaction-based Hardware Software Co-verification, Forum on specification & Design Languages, Darmstadt, 2006.

[6] D. Lettnin, M. Winterholer, A. Braun, J. Gerlach, J. Ruf, T. Kropf, W. Rosenstiel, Coverage Driven Verification Applied to Embedded Software; IEEE Symposium on VLSI, Porto Allegre, 2007.

[7] M. Winterholer, Verification of Embedded Software IP; IP'08 Grenoble, France, 2008.

[8] M. Winterholer, Metric Driven Test of Embedded Software, IP/ESC Conference, Grenoble, France, 2009.

[9] M. Winterholer, Metrik basierte Validierung und Verifikation von Software für Eingebettete Systeme, Entwicklerforum Embedded-System-Entwicklung, Ludwigsburg, Germany, 2009.

[10] M. Winterholer, Metric-Driven Validation and Verification of Software for Embedded Systems, Zuverlässigkeit und Entwurf, Stuttgart, Germany, 2009.

[11] M. Winterholer, Cadence System and Software Debug Capabilities, S4D Conference, Sophia Antipolis, France, 2009.

[12] D. Lettnin, M. Winterholer, Power Management Test and Verification for Hardware/Software Systems Embedded World Conference, Nuremberg, Germany, 2012.

[13] S. Brown, Hardware/software Verification with Incisive Software Extensions, Technical paper, 2011.

SCALABLE AND RETARGETABLE DEBUGGER ARCHITECTURE FOR HETEROGENEOUS MPSOCS

Luis Gabriel Murillo, Julian Harnath, Rainer Leupers, Gerd Ascheid

Institute for Communication Technologies and Embedded Systems (ICE)
RWTH Aachen University
Aachen, Germany
{murillo,harnath,leupers,ascheid}@ice.rwth-aachen.de

ABSTRACT

The increasing heterogeneity and parallelism of modern multi-processor systems on chip (MPSoCs) demand the evolution of existing debuggers in order to keep software development feasible. Such evolution will only be granted if upcoming software debuggers address key issues like abstraction, retargetability, scalability and convergence of information from different data sources. This paper presents a novel component-based, tree-aggregated debugger architecture which (i) grants flexibility and retargetability to deal with heterogeneous MPSoCs, and (ii) provides a framework to abstract complex details in order to facilitate debug of concurrency bugs. The debugger architecture is evaluated on an industrial-strength MPSoC virtual platform for mobile computing and next generation wireless communications.

Index Terms— Debugging, Parallel programming, Embedded software, Multicore processing, Virtual prototyping

1. INTRODUCTION

As embedded systems designers continue to include more parallel computational units in the same die, software debuggers must play an increasing role to keep programming complexity at manageable levels. The high degree of specialization required in modern MPSoCs results in a diverse combination of devices, programming models and software stacks, all coexisting in a single system. Such *cross-domain heterogeneity* from hardware to software creates several difficulties not only to inspect and analyze the execution of modern embedded software, but also to create a debugger architecture capable to cover and target all components in a system. From the hardware perspective, processing units as diverse as general-purpose processors, digital signal processors (DSPs) and application specific instruction set processors (ASIPs) offer different debug facilities with different access interfaces. From the software perspective, the coexistence of several complex software stacks demands a certain degree of awareness to abstract away irrelevant details and present only important data to the user.

Additionally, the parallelism of modern systems poses new challenges on the debug tools. Due to their inherent non-deterministic behavior, parallel applications are prone to concurrency bugs (e.g. data races and deadlocks) which are caused by unintended interactions of multiple processing units. Thus, debuggers targeting parallel systems need to aggregate debug information from different sources and provide a system-wide view. This means, in practice, collecting and merging all relevant debug information in a central facility. The architecture of a parallel debugger needs to be built to handle huge amounts of information without compromising its scalability, for instance, by avoiding single data storage components which could become a bottleneck.

Although debuggers dealing with cross-domain heterogeneity and concurrency issues are highly needed in the embedded systems community, this topic has been marginally addressed before. In consequence, modern MPSoCs still rely on the use of traditional debuggers, which requires attaching several instances of the same debugger, or even different debuggers, to the cores. Stepping through, running and inspection of software is then done individually for each core by the developer, and the task of keeping an overview of the complete system state becomes overwhelming.

This paper proposes a new retargetable debugger architecture which abstracts complex inner details of MPSoCs and their software stacks, thus facilitating a holistic perspective during debug activities. To achieve so, our debugger relies on the combination of the following concepts:

- A flexible, retargetable, component-based back-end framework which allows to make the debugger aware of different hardware and software combinations.

- An event-based intermediate representation for debug data which abstracts away low-level processor activity and facilitates data mining tasks and analysis of inter-processor interactions.

- A centralized tree-like facility to aggregate debug information from different sources, which helps to mitigate scalability and performance issues.

The rest of this paper is organized as follows. Related works are discussed in Section 2. In Section 3, the properties sought in next generation debuggers are drawn with basis on the characteristics of modern MPSoCs and software. Section 4 introduces the proposed debugger architecture, which is later completed with the explanations of the event-based intermediate representation and the aggregation mechanism in Sections 5 and 6. In Section 7, the debugger architecture is ported and used with an industrial-strength MPSoC virtual platform for mobile computing. With this case study the debugger's impact on the simulation time is analyzed. Finally, conclusions and future work are discussed in Section 8.

2. RELATED WORK

In recent years, debug paradigms have evolved significantly driven by the broad penetration of massively-parallel, high-performance computers (HPC). Debuggers have progressively moved from basic state inspection tools to advanced applications handling multiple concurrent processes and aggregate distributed state.

The works in [1, 2] firstly introduced debugger architectures which can scale properly to deal with significant amounts of cores. Adding a tree topology and a message aggregation mechanism, the authors appointed the Intel Debugger (IDB) [3] and the Hewlett-Packard Ladebug Debugger [4] to manage systems with significant concurrency. This kind of architecture was able to correctly cope with scalability by removing linear connections to a centralized debugger entity. A newer approach, which was based on offline analysis and data mining, was proposed by the authors in [5]. In this case, debug events are collected locally in the computation nodes at runtime and then merged in a database after program termination, thus giving a holistic view while avoiding system disturbances and performance issues.

From the commercial perspective, debuggers like DDT [6] have emerged with specialized architectures and interfaces that deal with big parallel machines. Such debuggers allow, for instance, visualizing a running parallel application in ways that are more intuitive to software developers (e.g. graph of a distributed data array).

In spite of these developments, debuggers for multicore embedded systems still linger behind. Although multicore debuggers with a unified interface have been introduced recently (e.g. ARM's DS-5 [7]), they provide but a traditional mechanism to run, step, inspect and control cores manually, which will become unmanageable with growing number of cores [5]. Furthermore, these debuggers generally target a single family of devices and provide limited retargetability, scalability and means to generate comprehensive abstractions. The debugger architecture herewith proposed tries to address this issue by applying concepts for scalability and aggregation from HPC debuggers under the flexibility requirements and especial constraints imposed by modern heterogeneous MPSoCs.

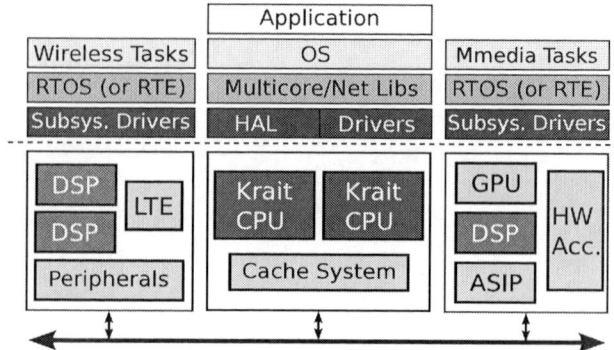

Fig. 1. Example of heterogeneity in a complex MPSoC - the Snapdragon (S4) MSM8960

3. NEXT-GENERATION MPSOC DEBUGGERS

This section presents the main characteristics of current and expected MPSoCs, which result in critical design questions from a debugger architecture's perspective.

3.1. Cross-domain Heterogeneity

Although embedded systems were traditionally designed to perform a specific function within a larger system, modern MPSoCs have surpassed this boundary by adding characteristics that resemble high-end general computing cores. Today's platforms often feature more than one application core, a fully-fledged OS, and are not restricted to a single specific function. Still, demands for power efficiency and throughput in multimedia and signal processing require specialized subsystems. Domain specific processors such as DSPs and ASIPs play an important role in embedded systems. These components add up to peripherals, accelerators and interconnects to create a complex heterogeneous hardware panorama which is difficult to program, integrate and verify from a system level perspective. An example of a modern commercial MPSoC, namely Qualcomm's Snapdragon S4 [8], is shown in Figure 1. This device consists of a dual-core application subsystem with ARM-like, OS capable Krait processors and two domain specific subsystems for wireless and multimedia with DSPs, GPU, ASIPs and hardware accelerators.

On the other hand, modern software has developed at a higher pace than hardware. For some application domains, the aforementioned trend to feature fully-fledged OSs is combined with firmware, middleware, specialized runtime environments (RTEs) and even real-time OSs (RTOSs). Multi-processing in such environments generally falls into the category of mixed *symmetric* and *asymmetric multi-processing* (SMP/AMP) because of differences in architectures, memory subsystems and address spaces. This has led to a scenario where a plethora of programming models coexist in a single device. In the platform shown in Figure 1, several software components can be highlighted:

- A main OS for general device control, resource administration and user-level applications which runs in the main subsystem.

- A software stack for real-time adaptability and multi-standard support which runs on the modem subsystem.

- Multithreaded RTEs or RTOSs which are executed in the DSP subsystems.

3.2. Debugging Concurrency

Anticipating how an application behaves in a non-deterministic concurrent MPSoC is difficult. In real applications, the effects of concurrency are usually underestimated, thus resulting in bugs like deadlocks, livelocks, atomicity and order violations that are neither found nor fixed easily [9].

Concurrency bugs are generally caused by a lack of correct synchronization, so the acts of synchronization need to be tracked. Expressive abstractions are useful to give the software developer a broad and generalized view of high-level synchronization software resources and their possibly faulty interactions. Abstracting the low-level architecture-dependent inner workings of a heterogeneous MPSoC to high-level state changes such as acts of synchronization is thus the key to efficient debugging. Bug-finding algorithms can then use the state change information to automatically detect potentially erroneous behavior.

Similarly, a debugger architecture must consider scalability issues in order to deal with increasing number of cores. On one hand, collecting huge amounts of data has a significant impact on the debugger's performance, and in extreme conditions, it can even exceed the resources of a single debug host machine. On the other hand, existing debug interfaces which provide a unified inspection mechanism become unmanageable with growing number of cores. Thus, a way to aggregate output from several data sources is necessary to condense the debug output into a more manageable subset.

3.3. Requirements for an MPSoC Debugger Architecture

Based on the previous analysis, a debugger targeting modern systems needs to be (i) aware of the differences that could appear both at the architectural and software levels and (ii) capable to handle complex concurrent systems were cores can possibly appear in hundreds or thousands. Thus, new MPSoC debuggers must be built on the following grounds:

- **Retargetability:** The architecture must facilitate adaptation to different underlying processing elements, while avoiding as much debugger source code manipulation or recompilation as possible.

- **Flexible Awareness:** A framework to setup software awareness must be provided as flexible as possible in order to address different coexisting software stacks

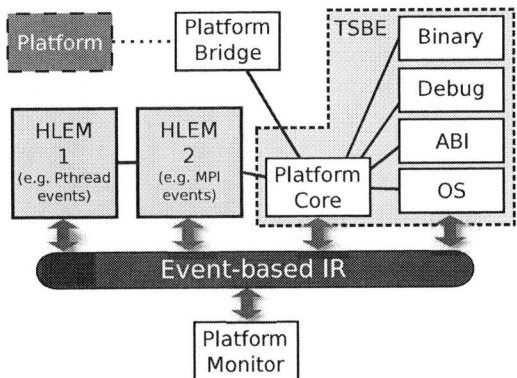

Fig. 2. Proposed component-based debugger architecture

and different granularity levels, ranging from bare-metal code to fully-fledged OSs.

- **Abstraction:** The debugger must rise the level of abstraction of the debug information in order to focus the user's attention on relevant system-level state changes instead of low-level details.

- **Scalability:** All the internal debugger components must be created and interconnected to support scaling up to hundreds or thousands of cores. Debug data filtering and aggregation are necessary to reduce bug analysis time and complexity.

4. PROPOSED ARCHITECTURE

The debugger architecture, shown in Figure 2, can be seen as a component-based design and is divided into components that communicate through well defined *interfaces* and an event-based intermediate representation (EIR). Components can be added, removed and replaced at any time depending on the target platform and the desired use case, thus providing high flexibility and extensibility. The main components behind the overall architecture can be subdivided as follows:

- **Platform Bridge.** To monitor an MPSoC's actions, the first step is to establish a communication channel for exchanging data and commands between the platform and the debugger. The exact nature of this channel is highly dependent on the platform itself (e.g. inspection APIs in virtual platforms, physical debug ports like ARM CoreSight DAP[10]). The *Platform Bridge* component is the first component which directly connects to the platform. After connecting, it abstracts all the communication details into a simple unified interface which is used by the rest of the framework.

- **Target-specific Back-End (TSBE).** The target-specific back-end is a set of components which communicate with each other and with the Platform Bridge in order to perform a specific control or inspection task for a given core. Breakpoint and watchpoint management, access to registers and memory inspection are some of the basic debug tasks which require direct access to a given physical resource. These basic tasks are handled by the *Platform Core* subcomponent which is directly connected to the Platform Bridge. Other debug tasks, such as understanding function prologues and epilogues, do not require direct access to a physical resource and can be implemented on top of the basic support given by the Platform Core. Thus, other target-specific components are introduced to manage the Application Binary Interface (*ABI*) of a core, the binary format of its executables (*Binary*), and the format used for its debug information (*Debug*). An *OS* component is also part of the TSBE because it is not uncommon anymore to see several OSs in a single platform, each targeting a specific core or a set of cores. The *OS* component deals with task structures, scheduling, virtual memory management and dynamic loading.

- **Intermediate Representation (EIR).** Apart from predefined interfaces that characterize particular components, each component can generate and react to *events*. Events act as an intermediate representation of debug information that abstracts low-level details of the platform and keeps only relevant information for the current debug task. All events in a system have a unified signature that allows debugger components to interoperate. Events can be either actions close to a core's hardware (i.e. a breakpoint hit, a program counter change, or a memory access) or programmer level actions which are composed of a series of low-level events (e.g. a function call, a task management operation, the execution of a synchronization primitive). The debugger IR will be described in Section 5.

- **High-level Event Monitors (HLEMs).** Programmer level events are key to abstract the complex inner workings of MPSoCs and give a comprehensive view of a concurrent system. Such events comprise API calls used explicitly by programmers to keep synchronization (e.g. a mutex lock) and dynamic OS actions that alter the tasks processing order (e.g. a task preemption event). Regardless of their complexity, awareness to these events can generally be added by using state machines with breakpoints, watchpoints and inspection functions on the target OS routines and structures (i.e. by using low-level IR events and APIs from the TSBE). Our framework provides the infrastructure to create event monitors which react to TSBE IR events and query the system state in order to condense more sensible information for the debug task.

- **Platform Monitor and Aggregation Tree.** Events do not flow directly between components. Instead, they are collected and distributed by the *Platform Monitor* component using a subscription mechanism. The Platform Monitor is also responsible for communicating events to analysis and data mining tools outside of the framework. Since a single dispatcher entity could become a bottleneck with increasing number of cores, a set of *Aggregation Interfaces* allow partitioning the debugger into component subsets to create a tree-like network of components. Such an architecture addresses scalability issues as explained in Section 6.

5. EVENT-BASED INTERMEDIATE REPRESENTATION

The debugger's EIR serves (i) as a binding mechanism for the debugger components and (ii) as an extensible ground layer to construct debug abstractions. The EIR is based on events which have a unified signature regardless of the granularity, the type, and the source. TSBE components and HLEMs act as producers and consumers of events, thus cooperating during a debug task to inspect and report system states that are critical for keeping correctness in a concurrent environment.

An event e describes a core's action together with the circumstances in which such action was performed. It is defined as a tuple $e = (t, a, p, c, d)$, consisting of a timestamp t, the type a of the performed action, the processing entity p which performed it, the execution context c which was active at the time and a set d with additional data specific to the event type.

Debug events are classified as *low-level* (EIR-LL) and *high-level* (EIR-HL) *events*. On one hand, EIR-LL events describe program actions happening close to the hardware. Events such as program counter changes and memory accesses are of particular importance because they define the application trace. The basic EIR-LL actions are *breakpoint* and *watchpoint hits*, and *memory read* and *memory write accesses*. Thus, the action type is defined as $a = \{hit_{bp}, hit_{wp}, mem_{write}, mem_{read}\}$. Depending on the system's complexity and the debug use case, other hardware related events which affect the task processing order and the system's state coherency, such as an MMU's page fault or an incoming interrupt, can be added to the EIR-LL set.

On the other hand, EIR-HL are related to complex software events that are important to the programmer, such as the creation of a task by the operating system or the successful acquisition of a synchronization resource. EIR-HL events are generally obtained after tracing the execution of OS and library functions which enable task management and multicore synchronization. Regardless of the actual implementation, actions commonly performed by multicore OSs and RTEs that are interesting while debugging can be identified and classified as shown in Table 1.

HLEM components generate EIR-HL events by putting together data coming from several EIR-LL events. Therefore, a HLEM is a system dependent implementation of a state

Class	Action Type (a)	Event Specific Data (d)
Function	$function_{call}$ $function_{return}$	name, parameters and return value
OS and Tasks	$task_{created}$ $task_{finished}$ $task_{scheduled}$ $task_{joined}$ $load_{dynamic}$	task ID, name, memory info, priority and status
Synchronization	$lock_{acquired}$ $lock_{released}$ $condvar_{wait}$ $condvar_{signal}$ $barrier_{wait}$ $barrier_{released}$	resource (lock, condvar, barrier) ID
Communication	$message_{send}$ $message_{receive}$	destination, source and data

Table 1. EIR-HL for common actions in OSs and RTEs

machine that consumes EIR-LL events and use them as state transitions. In every state, OS structures, memory and register values are obtained through the TSBE components' inspection APIs. Figure 3 shows the diagram of a POSIX threads (pthreads) [11] HLEM that identifies mutex lock calls with 5 states and 3 events. In this debugger architecture, the processing of EIR-HL events relies only on a breakpoint/watchpoint system, binaries compiled with debug data and state machines outside of the context of the target OS/RTE, thus avoiding the need for code instrumentation or other intrusive agents.

6. AGGREGATION INTERFACES AND AGGREGATION TREE

For hundreds or thousands of cores, the debugger performance will highly degrade if a centralized entity to manage inter-component event messaging is kept in the system. In this conditions, a single dispatcher will become a bottleneck with a complexity of $O(n)$ both for event dispatching and debug data collection, being n the number of nodes. Thus, instead of having a single *Platform Monitor*, a set of *Aggregation Interfaces* that interconnect *Platform Monitors* serve to create a tree-like network of components, as shown in Figure 4. Each *Platform Monitor* then connects to a single, or a reduced group of, target-specific back-ends and event monitors. The aggregation tree enables the parallelization of the debugger itself, and allows it to be executed as a distributed application when debugging a big system, while reducing the complexity of event dispatching and data collection to $O(log(n))$. Additionally, aggregation interfaces can be extended to condense the output of similar target-specific back-ends into a reduced and more manageable format, similarly to what proposed in [2]. This traduces into better debugger performance and higher debug efficiency from the user's perspective.

7. CASE STUDY

The proposed debugger architecture was implemented to target an MPSoC for mobile computing. The system is composed of a main processing subsystem with four ARM Cor-

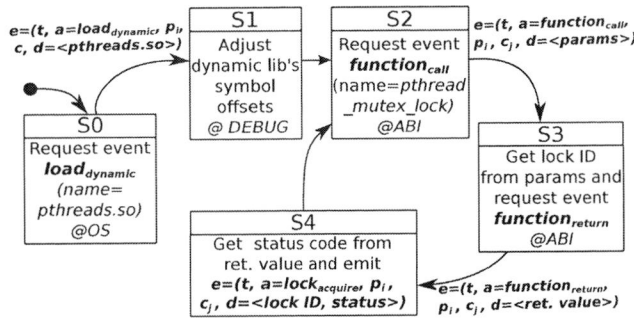

Fig. 3. Example of a HLEM for Pthreads mutex locks

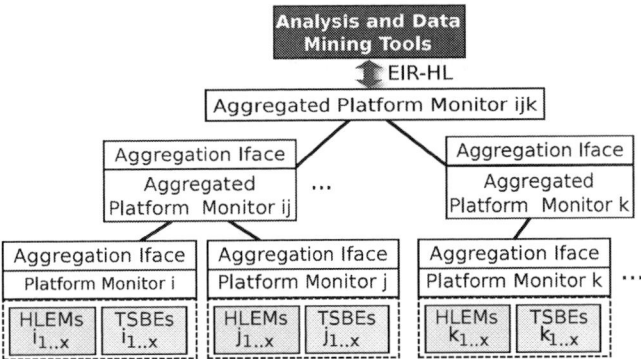

Fig. 4. Aggregated Platform Monitor tree

tex A9 cores [12] and a DSP subsystem with three Tensilica Xtensa cores [13]. The main processing subsystem executes a linux 2.6.35.9 kernel with SMP support. The DSP subsystem is devoted to wireless communications tasks and executes an OFDM transceiver on top of an in-house OS. Multicore synchronization and communication are supported through (i) pthreads in the main processing subsystem and (ii) the Xtensa Multiprocessing library (XMP) and mailbox based communication in the DSP subsystem.

The TSBE components were implemented for the ARM ABI, the Tensilica ABI, ELF object files, the DWARF2 debug format, linux and the DSP OS. The OS monitors generate EIR-HL events of the *OS and Tasks* class from Table 1 with state machines that trace kernel functions and structures, such as linux's `copy_process()`, `vma_link()` and `task_struct`. Due to the complexity of linux's memory and task management, the implementation of the TSBE took a considerable effort of approx. 1.5 man-months. HLEMs for pthreads and XMP were created to allow capturing and inspecting multi-processing events. The HLEMs trace the execution of synchronization and communication APIs and emit events on threads, mutexes, barriers and conditional variables. The event monitors were created with a minimum effort (i.e. approx. 1 man-days) thanks to the debugger's incremental component-based architecture.

Since only a Synopsys Platform Architect (PA) [14] virtual platform of the system is currently available, the debugger platform bridge was implemented to use PA's APIs to con-

49

Application	Number of EIR-LL events	Number of EIR-HL events	Total number of events	Time without debugger (s)	Time with debugger (s)	Time factor (X)	Events ratio (events/s)
FFT	378	222	**600**	0.649	2.079	**3.20**	582.4
FMM (256 cells)	3346	6206	**9552**	4.898	10.941	**2.23**	683.1
FMM (2048 cells)	21499	20994	**42493**	51.999	85.711	**1.65**	413.5
LU (continuous)	2086	1902	**3988**	51.047	60.187	**1.18**	40.9
LU (non-continuous)	714	558	**1272**	1.332	2.927	**2.20**	536.0
Ocean (continuous)	29414	27711	**57125**	117.426	134.877	**1.15**	250.5
Ocean (non-continuous)	27045	26870	**53915**	165.679	180.527	**1.09**	163.2
Radix	680	451	**1131**	11.83	13.529	**1.14**	57.5
Water Nsquared	103896	19210	**123106**	77.24	156.567	**2.03**	1,345.1
Water spatial	24306	805	**25111**	76.452	90.355	**1.18**	317.9

Table 2. Benchmark results for SPLASH-2 suite

nect to the simulator. Although this facilitates the implementation of the *Platform Bridge*, performance requirements for the debugger are very stringent as it must avoid decreasing the virtual platform speed to the point of compromising its usability. To test the debugger's performance, benchmarks from the modified SPLASH-2 benchmark suite [15] were used. The simulation speed was measured with and without the debugger, and the number of EIR-LL and EIR-HL events were obtained for the benchmarks. The obtained (wall-clock) simulation time and the time ratio with and without debugger are shown in Table 2. From the results it can be seen that the debugger creates an overhead but does not slow down the simulation significantly for applications with balanced processing/synchronization, such as in LU (cont), Ocean, Radix and Water Spatial. The event ratio (i.e. total number of events over simulation time), which tells how many breakpoints and watchpoints are handled by the debugger's state machines, is shown in the table as an indication of such balance. For applications with extremely high event ratio the simulation speed degrades by approx. 2X. FFT is a special case because it takes a negligible amount of time to be executed and the debugger initialization dominates in time.

8. CONCLUSIONS

Current debuggers for embedded systems lack several features necessary to target modern heterogeneous MPSoCs. On one hand, debuggers need to build on an architecture that facilitates retargetability and flexibility, in order to cover several combinations of cores, software stacks, programming models and OSs in a single system. On the other hand, debuggers must consider scalability and abstraction requirements posed by concurrent systems. This paper proposed a component-based debugger architecture which features an intermediate representation aimed at abstraction and gives the possibility to create a scalable aggregated tree structure. These elements combined form a unique MPSoC debugger architecture which can deal with heterogeneity and concurrency correctly. When testing the architecture in a simulation-based debugging scenario for a complex MPSoC and a representative parallel benchmark, it was shown that the debugger's performance does not heavily impact on the simulation time.

Future work will focus on new analysis and data mining tools for complex inter-processor interactions that operate at the IR level proposed by this framework.

9. REFERENCES

[1] Chih-Ping Chen, "The parallel debugging architecture in the intel debugger," in *PaCT*, 2003.

[2] Susanne Balle, Bevin Brett, Chih-Ping Chen, and David LaFrance-Linden, "Extending a traditional debugger to debug massively parallel applications," *Journal of Parallel and Distributed Computing*, 2004.

[3] Intel Corporation, "Intel Debugger IDB overview," http://software.intel.com/en-us/articles/idb-linux.

[4] Hewlett-Packard Company, "Ladebug Debugger," http://h30097.www3.hp.com/dtk/ladebug_ov.html.

[5] Karl Lindekugel, Anthony DiGirolamo, and Dan Stanzione, "Architecture for an offline parallel debugger," *International Symposium on Parallel and Distributed Processing with Applications*, 2008.

[6] Allinea Software, "DDT Debugger," http://www.allinea.com/products/ddt.

[7] ARM, "DS-5 debugger," http://www.arm.com/ds-5.

[8] Qualcomm, "Snapdragon S4," www.qualcomm.com/products/snapdragon.

[9] S. Lu, S. Park, E. Seo, and Y. Zhou, "Learning from mistakes: a comprehensive study on real world concurrency bug characteristics," *ASPLOS*, 2008.

[10] ARM, "CoreSight Debug Access Port," http://infocenter.arm.com/help/index.jsp?topic=/com.arm.doc.set.coresig%ht/index.html.

[11] 2004 Edition IEEE Std. 1003.1, "Posix.1c, threads extensions," http://pubs.opengroup.org/onlinepubs/009695399/basedefs/pthread.h.html.

[12] ARM, "Cortex A9 processor," http://www.arm.com/products/processors/cortex-a.

[13] Tensilica, "Diamond and Xtensa processor families," http://www.tensilica.com.

[14] Synopsys, "Platform Architect," http://www.synopsys.com/systems/architecturedesign.

[15] "The Modified SPLASH-2 Benchmark Suite," http://www.capsl.udel.edu/splash/.

CO-SIMULATION FRAMEWORK FOR VARIATION ANALYSIS OF RADIO FREQUENCY TRANSCEIVERS

Sumit Adhikari, Florian Schupfer and Christoph Grimm

ICT, Vienna University of Technology
email: {adhikari,schupfer,grimm}@ict.tuwien.ac.at

ABSTRACT

Co-simulation provides an architect or a designer the freedom to change the implementation of a sub-module and analyse the design. Variational analysis during design procedure is an important tool to ensure higher-yield during production. State of the art co-simulators uses Monte Carlo analysis method for variational analysis, which is a multi-run method, slow and outcomes are not completely covered. In this article we proposed a co-simulation environment which uses affine arithmetic as variational analysis method. The result is an efficient and completely covered co-simulation environment.

Index Terms— Electronic System Level (ESL), Co-simulation Framework, Affine Arithmetic, Variation Analysis, Design Reliability, Abstract Modelling.

I. INTRODUCTION

A Simulation methodology called co-operative simulation or co-simulation allows sub-modules of a design to be simulated by different simulators by co-operative exchange of information during simulation. Co-simulation using coupled simulators provides many additional degrees of freedom to the architect or the designer. Especially in the case when many implementations of the same subjects are available as it has been described in [1] and shown in Figure 1. The designer will have scope to analyse many performances just by changing implementation of the model he or she is using.

Unfortunately these different implementation possibilities are not homogeneous, rather they are heterogeneous - written in different languages. There are many co-simulation environment available with many EDA/ESL vendors in order to perform co-simulation among various implementations written in different languages. Worst case behaviour analysis of a system is an important paradigm in analysing the production tolerance of the system. Variation analysis for the reliability of chosen parameters at circuit level has been efficiently analysed using *Monte Carlo (M.C.)* simulation through past many decades by analog designers. However, the quality of analysis does not ensure complete coverage of every corners for every parameters as the simulation relies on generation of random values. Furthermore, M.C. analysis relies on multi-run of simulation environment which demands precious simulation time, or in other words the method is slow. Authors of this article has exploited M.C. in the *Electronic System Level (ESL)* modelling and found it insufficient in terms of corner coverage and simulation speed. Many parameters influencing the design reliability leads to untested corners which might arise due to non-occurrence of randomness in the variation of parameters due to constrained (fixed) number of simulation run. On the other hand increase in simulation run reduces the chance of non-coverage by demanding simulation run-time. Even with a high number of simulation runs, the results does not ensure complete coverage of the parameter variation for obvious statistical reasons. With M.C. into place fully covered variation analysis stays as a problem in every level of electronic design-automation.

Instead of approaching the systems worst case behaviour by stochastic methods, semi-symbolic simulations using affine arithmetic compute a range based system response in one simulation run. Deviations of nominal system parameters are modelled as ranges and superimposed to the nominal system model. The mathematical concept of affine arithmetic [2] provides a definition for a set of deviation ranges and additionally specifies mathematical operations on these ranges. Variations of system parameters and quantities are considered by continuous ranges and a pessimistic but guaranteed range for the system response is found on either the system level or circuit level [3]. The range based system response can be used for an advanced analysis of the deviated system behaviour [4] or to identify refinement candidates to improve the reliability of the overall system [5]. Affine Arithmetic labels the ranges by symbols which allows the instant backtracking of the system response contributions to their sources. Although the most published work concentrates on the system level also transistor level circuits are simulated when solving the non-linear differential equations by using Affine Arithmetic [6]. [7] uses semi-symbolic simulation to analyse the convergence behaviour of control loops in presence of uncertainties. [8] and [9] additionally enhance the semi-symbolic simulation for simulating non-linear analog circuits and obtaining refinement information to improve the system quality. The problem of over-approximation is addressed in several works, where the Affine Arithmetic is enhanced by additional affine operations which result in exact solutions. [10]–[12]

I-A. Affine Arithmetic

Affine Arithmetic is a methodology to define ranges as super-position of a center value with a set of $\mathcal{N}_{\tilde{x}}$ sub-ranges. Affine Arithmetic bases on the original concept of Interval Arithmetic [13] but enhances it with symbolic range identifiers to overcome the dependency problem preventing the usability of the Interval Arithmetic. Along with the formulation of mathematical operations on these ranges an arithmetic is established [2]. Uncertainties in systems are modelled as ranges and these ranges are modelled by so called Affine Forms. Mathematical calculations on Affine Forms are defined and provide at least pessimistic approximations to allow worst case evaluations. $\mathcal{N}_{\tilde{x}}$ defines a set of natural numbers identifying all deviation terms $x_i \epsilon_i$ in symbol $\tilde{x}$.

$$\tilde{x} = x_0 + \sum_{i \in \mathcal{N}_{\tilde{x}}} x_i \epsilon_i \qquad \epsilon_i \in [-1, 1] \qquad (1)$$

The mathematical operations can be divided into two classes, affine operations which solve in exact results and non-affine operations

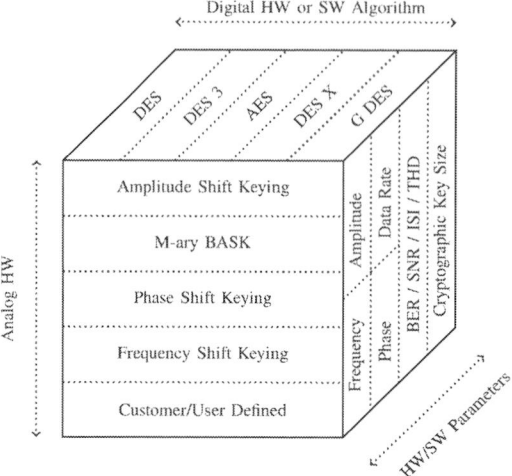

Fig. 1. An Architect's Cube of Possibilities

which are derived as pessimistic approximations. Pessimistic approximations are considered to safely contain the actual operation result but over-approximates the range since the exact result can not be formally determined. Affine operations are the addition and subtraction of Affine Forms as well as the multiplication of Affine Forms by numeric values defined by:

$$\tilde{x} \pm \tilde{y} = (x_0 \pm y_0) + \sum_{i \in \mathcal{N}_{\tilde{x}}} (x_i \pm y_i)\epsilon_i \qquad (2)$$

$$c\tilde{x} = cx_0 + \sum_{i \in \mathcal{N}_{\tilde{x}}} cx_i\epsilon_i \qquad (3)$$

Non-affine operations are derived by an approximation of the resulting Affine Form. Since the approximation is considered as being pessimistic, non-affine operations are the source of over-approximations which influences the simulation expressiveness negatively. Significant over-approximation can prohibit the usage of semi-symbolic simulations as the system behavior is concealed by wide ranges. Improvements on Affine Arithmetic have been introduced in recent years to reduce the over-approximation effects. For instance the Quadratic Arithmetic [6], [12] has been introduced which adds multiplications and the square function of Affine Forms to the affine operations which now also produce exact results.

I-B. SystemC AMS

SystemC AMS uses C++ based language constructs to model and simulate analog and/or mixed-signal systems [14]. Its main *Model of Computation* (MoC) is *Timed Synchronous Data-flow* (TDF) which is also used for the semi-symbolic simulation. It is a timed version of the original *Synchronous Data-flow* (SDF) which allows to pre-calculate the schedule of process executions. This characteristic offers a high simulation performance in combination with a powerful modelling expressiveness. On the other hand the C++ based nature of SystemC AMS allows easy integration of additional libraries, like the Affine Arithmetic library for semi-symbolic simulations. This extensibility makes SystemC AMS an

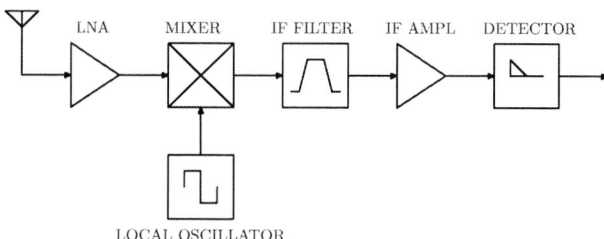

Fig. 2. Schematic of Radio Frequency Receiver

efficient choice for the ever increasing functionality of the range based approach.

Many co-simulation tools are commercially and academically available which does co-simulation and variational analysis using Monte Carlo analysis. There is no literature available at this point which shows availability of a co-simulation environment which supports variational analysis using Affine Arithmetic. This article addresses this issue. In this article we propose a co-simulation environment which supports heterogeneous transient simulation and variational analysis between a circuit level affine simulator and SystemC/SystemC AMS simulator using Affine Arithmetic. As an use case here we use a superheterodyne receiver on which we perform co-simulation between a circuit level affine simulator and SystemC AMS simulator. We start with describing the design, then discuss a refinement and show our simulation results. Finally we conclude our article.

II. DESCRIPTION OF THE DESIGN

The system under examination is a superheterodyne receiver as shown in Figure 2. The small-amplitude signal is first picked by an antenna which is then amplified by a *Low Noise Amplifier (LNA)*. The signal is then mixed with the frequency of a *Local Oscillator (LO)*. Idealistically at this stage the mixer output should generate the desired output. But due to the fact that there will be always

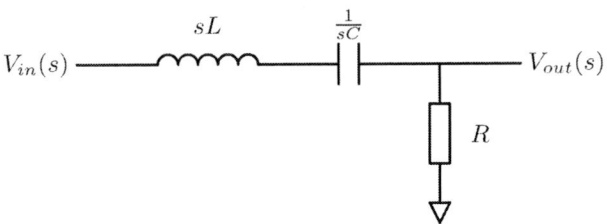

Fig. 3. Schematic of IF Filter

a phase shift of the LO frequency with the transmitter oscillator frequency plus the non-linearity of LNA and mixer, the output of the mixer will always contain multiple frequencies which are called image frequencies. These image frequencies can be removed by a band-pass filter called IF-Filter. The signal is then amplified using an IF-Amplifier and then detected and passed over to following stages. In our current context a fundamental signal $F = 0.5\text{MHz}$ is mixed with a frequency $F_{OSC} = 10\text{MHz}$. In the case of the mixer at the receiver side, this will generate multiple frequency of F with combination of F_{OSC}. These various frequencies are called image frequencies and can be effectively filtered out using a gentle bandpass filter.

II-A. Design of the IF-Filter

In this case we consider a passive LC bandpass filter in which the following stage is an active IF amplifier and hence the LC bandpass will always see an ideally infinite (very high in reality) input impedance. The filter under consideration is shown in Figure 3, in which the inductor L is connected in series with a capacitor C and resistor R. This is a series LCR circuit, where the output is measured across the resistor R. The transfer function of such a circuit is,

$$H(s) = \frac{sRC}{1 + sRC + s^2 LC} \tag{4}$$

The circuit in Figure 3 should have a resonance at $\omega_0 = 2\pi F$ for it to reject the image frequencies due to mixer efficiently. The

complex frequency response of Equation 4 is given by,

$$H(\omega) = \frac{R}{R + j\left(\omega L - \dfrac{1}{\omega C}\right)} \tag{5}$$

From Equation 5, the resonance frequency can be found out as,

$$f_0 = \frac{\omega_0}{2\pi} = \frac{1}{2\pi\sqrt{LC}} \tag{6}$$

The quality factor of the filter is defined as,

$$Q = \sqrt{\frac{L}{R^2 C}} \tag{7}$$

With proper choices of L, C and R, the frequency response of the IF filter has been shown in Figure 4 which shows that the filter under consideration is indeed a bandpass. The quality of the IF filter has been shown in Figure 5 which guarantees rejection of unwanted components by approximately 60dB, which ensures 10-bit resolution under noiseless idealistic conditions with all preceding stages idealistic.

II-B. Choice of Oscillator Waveform

Design of sinusoidal waveform out of an oscillator is generally difficult and expensive. Whereas, an oscillator waveform shown in Figure 6, which has significantly high transition times are easy to design and definitely consumes less current from power supplies. The aspiration of using such a oscillator waveform is based on assumption that, in reality, the mixer is an extremely non-linear device. This non-linearity when acts on oscillator waveform like as shown in Figure 6, the output of the mixer appears as if it has been mixed with a perfect sinusoidal oscillator waveform. The co-simulation will also validate the correctness of the choice of the oscillator waveform.

III. REFINEMENT OF THE DESIGN

When all the constituent blocks are idealistic, it is impossible to predict the performance of the design under consideration. Hence,

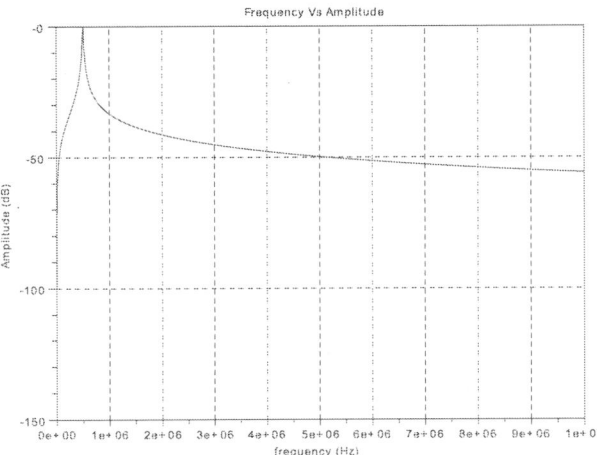

Fig. 4. LC Filter Transfer Function

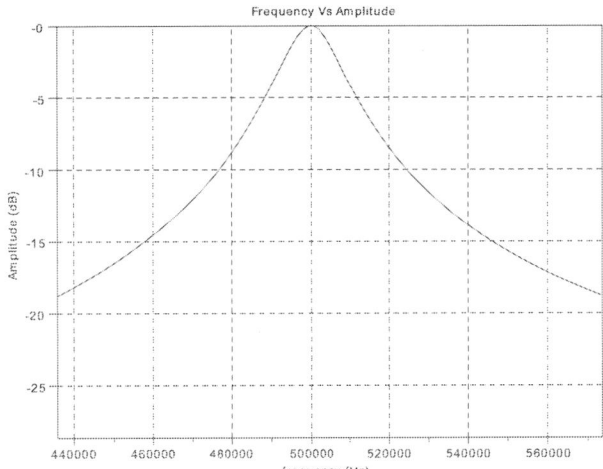

Fig. 5. LC Filter Transfer Function (Zoomed)

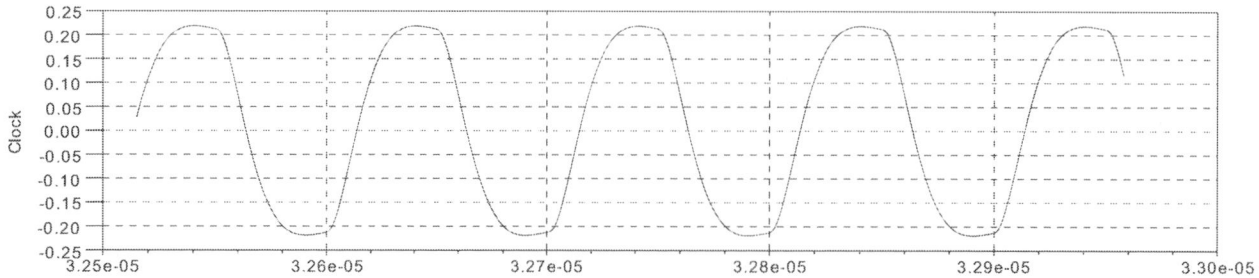

Fig. 6. Schematic of Oscillator Output

hence idealistic SystemC AMS blocks of Figure 2 needs to be refined one at a time in order to do a performance analysis on the architecture under consideration. The idea is to systematically replace functional block one by one starting from the beginning of the signal chain with circuit level blocks keeping all other blocks at architectural level and validate the performance of the design. Before we proceed, we discuss the simulation environment first in our upcoming Section.

III-A. Description of Heterogeneous Simulation Environment

The simulation environment uses co-simulation between the SystemC AMS simulator and the range-based analog simulation environment as described in [15], [16]. Generally, a co-simulation environment is a collection of simulation engine, certain signal representation and a collection of models written in different languages. Heterogeneous signal representations are seamlessly interfaced which poses a need of converter interfaces. In general a complicated backplane is needed to manage the synchronization between models, interfaces and engines. In our case we assumed that the scheduling is static as in SystemC AMS simulator which alleviates the need to additional synchronization and hence alleviates need of an additional complicated simulation backplane. This essentially speeds up the execution speed which is absolutely necessary at this level of simulation. The analog simulator is wrapped using a SystemC wrapper which handles all the coupling between the two simulators. Any SystemC AMS signal can be interfaced with the SystemC wrapper over the analog simulator. Whenever an event is created by the SystemC wrapper, the analog simulator is called and provided with the new value. The analog simulator accepts this updated value and runs the simulation. In between two successive SystemC events, the analog simulator might generate many outputs which are eventually rejected by the coupling mechanism. The analog simulator generates a nominal value output and a range output which can be used by the SystemC AMS simulator to construct affine signals for post processing. The analog simulator used is a linear ordinary differential equation solver which uses backward Euler method to compute solutions at desired time-steps. Thus, the accuracy of the analog solver is heavily dependent of the sampling time of the input data provided by SystemC AMS simulator. Hence, it is important to analyse and set the sampling time of the SystemC AMS simulator before the refinement procedure starts. It is also important to mention that the analog simulator although generates affine outputs but it does not accepts affine inputs. Hence, it is always necessary for preceding blocks to generate non-affine signals.

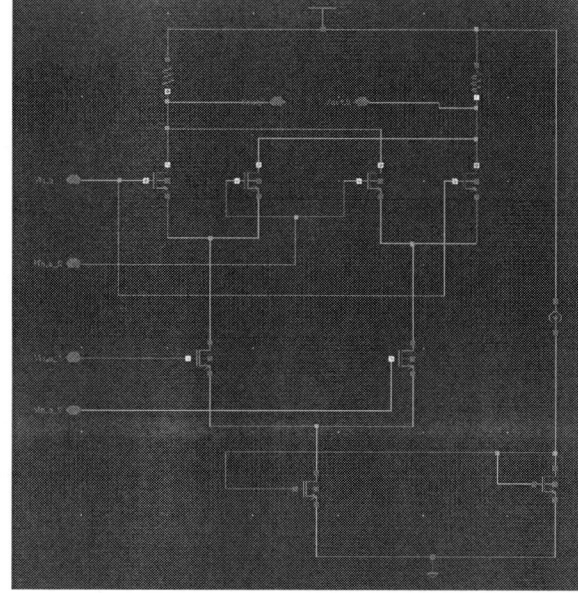

Fig. 7. Schematic of Circuit Level Mixer

III-B. Refinement Case

Consider the case (which is one out of many possibilities) in which we have a mixer implemented in circuit level. We replace the SystemC AMS mixer block by a circuit level mixer of Figure 7. The circuit in Figure 7 is a resistor connected differential cross-coupled mixer which has been implement using MOS transistors. Section IV presents and discusses the results of this refinement-case in detail.

IV. SIMULATION RESULTS

With the mixed-level environment, when we performed the mixed-mode simulation, we found that the simulation is 100X slower than the pure SystemC AMS simulation, which is completely justified for the current purpose considering the gain from the simulation.

The nominal value simulation results are shown in Figure 8. The clock is mixed with a sinusoid and provided as the mixer input. The second sub-figure from top shows the mixer input. This input

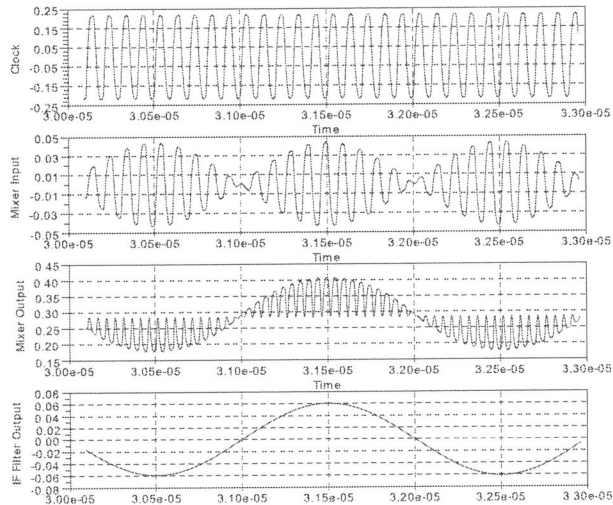

Fig. 8. Transient Simulation Result of the Receiver

when passed through the circuit level mixer, outputs the signal shown in the third sub-figure. Clearly, it contains image components and it appears as if it has been mixed with a sinusoidal oscillator frequency. The fourth sub-figure shows the output of the IF-filter, which is clean and free from image frequencies. It has to be noticed that due to the passive nature of the image filter the common-mode of it's output is disturbed.

The most important analysis at this point would be variational analysis. Considering every component in the IF filter has a manufacturing tolerance as 1%, Figure 9 shows the variational analysis result. The top sub-figure is the mixer output including all circuit level manufacturing tolerances. The bottom sub-figure, showing the output of IF-filter, has only 8% variation which is absolutely acceptable. This is to mention that actual variation would be less than 8% as the affine arithmetic operations suffers from over-approximation.

V. CONCLUSION

In this article we discussed the power of co-operative simulation or co-simulation for an architect or a designer. We also discussed that the variational analysis methods used by state of art co-simulators are Monte Carlo simulation method which is multi-run, slow and not completely covered. We proposed a co-simulation framework between a circuit-level simulator and SystemC/SystemC AMS simulation engine which supports variational analysis using Affine Arithmetic. We discussed the complete design of a superheterodyne RF receiver. We then discussed co-simulation the use-case of the RF receiver in which we replaced the mixer module with an alternative, more detailed circuit level mixer module. We then demonstrated the nominal value simulation result and the variational simulation result. The co-simulation framework, which has been described here supports affine analysis is a novel approach towards this kind of variational analysis. A SystemC AMS simulator which is statically scheduled, being the master of the co-simulation environment alleviates the need of a simulation backplane. Due to this reason the simulation interfaces

are simple, the simulation is extremely fast in comparison to general other simulation environment.

VI. FUTURE WORK

As a continuation of our current work, we would like to integrate more heterogeneous components which will be executed by other simulators (like Verilog HDL) and see how we can improve the performance of the simulation.

ACKNOWLEDGEMENT

This work has been has been funded by the Vienna Science and Technology Fund (WWTF) through project ICT08-012.

VII. REFERENCES

[1] Y. Zaidi, S. Adhikari, and C. Grimm, "Abstract Modeling and Simulation Based Selective Estimation," in *Design and Diagnostics of Electronic Circuits Systems (DDECS), 2011 IEEE 14th International Symposium on*, April 2011, pp. 275–278.

[2] L. H. de Figueiredo and F. Stolffi, *Self-Validated Numerical Methods and Applications*, ser. Brazilian Mathematics Colloquium monographs. IMPA/CNPq, Rio de Janeiro, Brazil, 1997.

[3] F. Schupfer and C. Grimm, "Towards more Dependable Verification of Mixed-Signal Systems," in *Verification over discrete-continuous boundaries*, ser. Dagstuhl Seminar Proceedings, no. 10271, 2010.

[4] F. Schupfer, M. Kargel, C. Grimm, M. Olbrich, and E. Barke, "Towards Abstract Analysis Techniques for Range Based System Simulations," in *Specification Design Languages, 2010. IC 2010. Forum on*, sept. 2010, pp. 1 –6.

[5] F. Schupfer, C. Radojicic, J. Wenninger, and C. Grimm, "System Refinement Design Flow based on Semi-Symbolic Simulations," in *AFRICON, 2011. AFRICON '10*, 2011, pp. 1–6.

[6] D. Grabowski, M. Olbrich, and E. Barke, "Analog circuit simulation using range arithmetics," in *ASP-DAC '08: Proceedings of the 2008 Asia and South Pacific Design Automation Conference*. Seoul, Korea: IEEE Computer Society Press, 2008, pp. 762–767.

[7] C. Grimm, W. Heupke, and K. Waldschmidt, "Analysis of Mixed-Signal Systems with Affine Arithmetic," *IEEE Transactions on Computer Aided Design of Circuits and Systems*, vol. 24, no. 1, pp. 118– 123, 2005.

[8] D. Grabowski, C. Grimm, and E. Barke, "Ein Verfahren zur Effizienten Analyse von Schaltungen mit Parametervariationen," in *Tagungsband GI/ITG/GMM - Workshop Modellierung und Verifikation '03*, Dresden, Mar 2006.

[9] C. Grimm, W. Heupke, and K. Waldschmidt, "Refinement of Mixed-Signal Systems with Affine Arithmetic," in *Design, Automation and Test in Europe 2004 (DATE 04)*. IEEE Press, Feb 2004.

[10] H. Shou, H. Lin, R. Martin, and G. Wang, "Modified Affine Arithmetic Is More Accurate than Centered Interval Arithmetic or Affine Arithmetic," in *Martin(Eds.), Lecture Notes in Computer Science 2768, Mathematics of Surfaces, Springer-Verlag*. Springer, 2003, pp. 355–365.

[11] F. Messine and A. Touhami, "A General Reliable Quadratic Form: An Extension of Affine Arithmetic," *Reliable Computing*, vol. 12, no. 3, pp. 171–192, 2006.

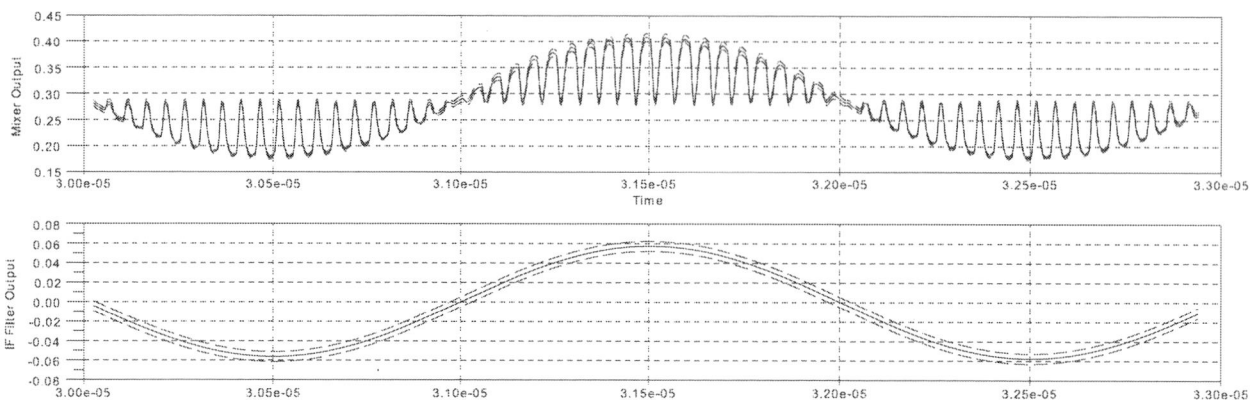

Fig. 9. Variational Analysis Result of the Receiver

[12] D. Grabowski, M. Olbrich, C. Grimm, and E. Barke, "Range Arithmetics to Speed up Reachability of Analog Systems," in *Forum on specification and Design Languages, FDL 2007*, 2007, pp. 38–43.

[13] R. E. Moore, *Interval Analysis*. Eaglewood Cliffs, NJ: Prentice-Hall, 1966.

[14] A. Vachoux, C. Grimm, and K. Einwich, "Analog and Mixed-Signal Modeling with SystemC-AMS," in *International Symposium on Circuits and Systems 2003 (ISCAS '03), IEEE Press, 2003*.

[15] E. Barke, D. Grabowski, H. Graeb, L. Hedrich, S. Heinen, R. Popp, S. Steinhorst, and Y. Wang, "Formal approaches to analog circuit verification," in *DATE*, 2009, pp. 724–729.

[16] D. Grabowski, M. Olbrich, and E. Barke, "Analog circuit simulation using range arithmetics," in *Design Automation Conference, 2008. ASPDAC 2008. Asia and South Pacific*, march 2008, pp. 762 –767.

Vienna, Austria – September 19-20, 2012

Session 4: Tracing

Compact Function Trace (CFT)
Albrecht Mayer and Reinhard Deml (Infineon)

CULT: A Unified Framework for Tracing and Logging C-based Designs
Wei Hong, Alexander Viehl (Forschungszentrum Informatik Karlsruhe), Nico Bannow, Christian Kerstan, Hendrik Post (Robert Bosch GmbH), Oliver Bringmann and Wolfgang Rosenstiel (University of Tuebingen)

ARMv8 Debug and Trace Architectures
Michael Williams, ARM Limited

www.ecsi.org/s4d

Compact Function Trace (CFT)

Albrecht Mayer, Reinhard Deml

Infineon Technologies AG

albrecht.mayer@infineon.com, reinhard.deml@infineon.com

Abstract

Unlimited program flow trace is used for performance analysis on function level and for debugging of sporadic effects. Trace output via an affordable and robust tool interface like Infineon's two pin DAP [1] requires a very high trace compression. With a novel approach the required compression is achieved by tracing just calls and returns of functions and by filtering out completely very short functions, which do not call any other function.

1. Introduction

Tracing of the program flow is used for debugging and performance analysis. For debugging usually only a limited trace depth is needed, if the trigger logic allows stopping the trace recording at or shortly after the error symptom. However for performance analysis long traces with high timing resolution are needed to measure the min/max/average function execution time. In addition this analysis can be flat or nested with the full hierarchy of function calls.

The following figure shows a typical representation of such trace data by a tool. There are numerous other representations which can represent the behavior over time or statistically.

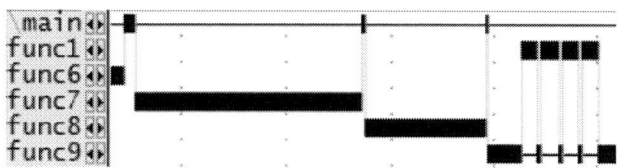

Figure 1 Function Flow Example

In current solutions the chip has a continuous trace interface like Nexus or ETM which allows one to record and time stamp the program flow externally. This approach has two disadvantages. The main disadvantage is the required high bandwidth interface for the trace. This means usually at least 5 trace pins. The other disadvantage is that the time stamping is done externally. Due to varying fill level of the on-chip FIFO for the trace messages, this results in an inaccuracy of the time stamps.

Newer solutions have a high-speed serial trace interface and can thus afford on-chip time stamping. However

such an interface leads to a significant chip area increase, special packages, a more costly PCB technology and challenging high-speed signal integrity system design requirements.

2. Statistical Analysis

A flat performance analysis at a function level for the high-runners can be easily achieved by just statistically sampling the program counter. The accuracy of such a measurement can be scaled with the number of samples either by faster sampling or by accumulating samples from different runs. For modern trace architectures this is possible with no or minimum run-time impact.

A nested performance analysis at a function level means that it is possible to determine the run time of a high-level function including all sub-functions on all levels. This requires time stamped recording of all function calls and returns, of all task switches and of all interrupt service routines. If parts of this information are missing (e.g. a task switch) due to a trace interface overrun, the complete statistical analysis becomes impossible or at least questionable due to potentially wrong assumptions in the recovery from the overrun.

If the chip has an on-chip trace buffer and the setup for the statistical analysis is non-hard realtime, the trace data collection can be split into fractions which fit into the buffer [2]. However if the system requires hard realtime behavior, such an approach is not possible.

3. Debugging

Also in certain debug scenarios an unlimited program flow trace can be helpful in particular when data accesses to specific locations e.g. a peripheral are being traced. A typical application is the analysis of overload situations, where interrupts and higher priority tasks result in missed deadlines. Such debug cases can of course occur in a non-lab, final product and target system setup, where it would be not possible to connect a high-speed trace interface.

4. Compact Function Trace (CFT)

For the analysis and debugging purposes described before it is necessary to trace calls and returns of functions and interrupt service routine and data associated with task switches.

Compact function trace means that the output of trace messages for short leaf functions is reduced or even omitted. The term "leaf function" is used here for functions in the flow tree which do not call any other functions. Please note that the latter is not necessarily a static property of a function.

Briefly described the CFT outputs for each call and return the source address, the target address and a time stamp. Overall this amounts to about 150 bits. The addresses are often not very distant, so the usual address data compression schemes reduce this data significantly. Depending on robustness and compression considerations, it is even possible to omit the output of target addresses in certain situations.

The CFT rules are:

- For very short leaf functions no trace at all is generated (omitted functions)
- For short leaf functions only the call instruction outputs a trace message

The function execution length threshold for these decisions is configurable and could be potentially also controlled by the trace buffer fill level.

This high level function trace can be enriched with statistical performance analysis data like IPC (instructions per cycle) information [5]. Since the IPC measurement is anyway averaging over e.g. 100 clock cycles it is a very good match with the function level granularity of CFT.

For debugging purposes there are two extensions:

- If there is a data access being traced, call/return messages are always generated for the leaf function where it occurs
- The trace resolution can be changed with a trigger. E.g. with the entry to a "function of interest" it is changed from CFT to a high resolution trace mode.

If a high level trace with even further reduced bandwidth is good enough, it would be possible to extend the idea to short non-leaf functions which call (very) short leaf function(s).

5. Automotive Powertrain Use Case Example

A statistical analysis was made for different powertrain application traces. For the most challenging one, there was on average a function call every 60 CPU clock cycles. 80% of the function calls were to leaf functions. 22% of all function calls took 8 cycles or less. 39% took 16 cycles or less. 15% call again the same function as just before.

If the threshold for omitted functions is 16 cycles and for "only call trace" leaf functions is 100 cycles, then on average only 45 bits of trace data are generated for a function (rough estimation). This means for a 100 MHz CPU less than 10 MByte/s of trace data is generated, which is less than the available bandwidth of advanced two pin tool interfaces.

6. Conclusions

CFT provides an unlimited trace of a processor core, running at more than 100 MHz over a regular two pin tool interface. This requires no special packages and system design and is affordable not only for emulation devices and development ECUs, but also for mass production devices and products. In combination with a high resolution on-chip trace for zooming into short sequences, this covers the full range for debugging and performance analysis in a very cost effective way.

7. References

[1] Infineon's 32-Bit Microcontroller documentation http://www.infineon.com Product category: Microcontrollers

[2] A. Mayer, "Method for emulating an integrated circuit and semiconductor chip for practicing the method", German Patent DE102004008499, US Patent Application US2005192791

[3] A. Mayer, H. Siebert, K. D. McDonald-Maier, "Boosting Debugging Support for Complex Systems on Chip." IEEE Computer Magazine, April 2007

[4] Nexus standard http://www.nexus5001.org/standard

[5] A. Mayer, F. Hellwig, "System Performance Optimization Methodology for Infineon's 32-Bit Automotive Microcontroller Architecture", http://www.date-conference.com/proceedings/PAPERS/2008/DATE08/PDFFILES/08.2_3.PDF

CULT: A Unified Framework for Tracing and Logging C-based Designs

Wei Hong[1], Alexander Viehl[1], Nico Bannow[2], Christian Kerstan[2], Hendrik Post[2]
Oliver Bringmann[1,3], Wolfgang Rosenstiel[1,3]

[1]FZI Forschungszentrum Informatik	[2]Robert Bosch GmbH	[3]Universität Tübingen
Haid-und-Neu-Str.10-14	Robert-Bosch-Str.2	Sand 13
76131 Karlsruhe, Germany	71701 Schwieberdingen, Germany	72076 Tübingen, Germany

[1][hong,viehl]@fzi.de
[2][Nico.Bannow,Christian.Kerstan,Hendrik.Post]@de.bosch.com
[3][bringman,rosenstiel]@informatik.uni-tuebingen.de

Abstract—**This paper presents a novel framework for tracing and logging C-based designs of embedded hardware/software systems. The development of this C-based Unified Logging and Tracing (CULT) framework was driven by the necessity of a common development support environment between different design disciplines and at different early design phases in which development takes place using C-based languages (C/C++/SystemC). The elaborated framework is highly configurable and scalable. It requires only minimal code changes in order to be used and it implements enhanced features as a dynamic configurable history and backtracking of signals and values. Further, we demonstrate the interconnection with assertion-based verification (ABV) in order to support design testing and validation in early design phases. The framework will be released as open-source project for establishing a vital community and for building up a tooling landscape around the basic capabilities.**

Index Terms—**Logging, Tracing, SystemC, SystemC TLM, SystemC AMS**

I. INTRODUCTION

SystemC [1] is a widely used simulation library for system-level modelling, architectural exploration and other fields of application. It provides an event-driven simulation kernel, based on C++ classes and macros. After its first release a decade ago, SystemC has taken numerous changes, but the native SystemC logging and tracing support is still insufficient for the diagnosis and debugging of complex systems. Furthermore, the introduction of new features, such as Transaction-Level Modelling (TLM) [2] and Analog and Mixed-Signal (AMS) [3] extensions, also require the support of a powerful logging and tracing mechanisms. With TLM, details of the communication among the individual modules are abstracted and separated from the communication architecture. Hence, complex and cross-module transactions are abstracted from the actual realization in hardware and/or software. During the simulation of the models, it is important to observe data concerning functional and non-functional properties. The native SystemC logging and tracing mechanism traces the value of a signal or variable with a sc_trace() method. It stores the records as VCD files, which does not provide sufficient flexibility and functionality. Current logging and tracing frameworks are not flexible or powerful enough to satisfy the specific logging and tracing requirements of complex systems

under design. The tracing library [4] from LogicPoet for example is only specialized for logging events or event-based transactions. To meet the requirement of an unified interface and library for the efficient recording of simulation data, we introduce CULT, which stands for a C-based Unified Logging and Tracing framework.

CULT provides an efficient logging and tracing framework implemented as a static C++ library. It targets the configurable and customizable recording of simulation data for different abstraction levels. Hence, not only C++ applications are supported by the framework, but also SystemC applications and its TLM and AMS extensions. Furthermore, it is extensible by a layered architecture approach. The most important properties of CULT are:

1) logging and tracing support of different object types: standard C++ data types, SystemC signals, SystemC ports, TLM transactions and SystemC AMS signals,
2) flexible dynamic configuration,
3) conditional logging,
4) multiple output format of the logging and tracing data,
5) minimal-invasive intervention of the original code,
6) interconnection with assertion-based verification [5].

Technical background information will be introduced in Section II. In section III, CULT will be presented in detail. Section IV describes the application and performance impact of the framework. The last Section V concludes the paper and concerns future development aspects and ideas of the framework.

II. PRELIMINARIES

A. SystemC

The time-to-market demands and hence the length of the development cycle of embedded systems has become shorter from year to year. This is because of the strong competition, which forces the companies to bring their new products faster to the market. But short design cycles demand early prototyping to achieve parallel design of the system's software and hardware. This will help to find and fix most bugs at a very early design stage, which results in lower costs for design iterations. SystemC is a modelling language on system-level, which supports this purpose. Basically, it provides a C++ library for modelling concurrent systems, which has an event-driven simulation kernel. It supports several abstractions from pure functional to cycle accurate level. Currently, SystemC is emerging as one of the premier design language standards.

B. Transaction-Level Modelling

Transaction-level modelling is a high level approach of modelling electronic systems. Details of the functionality among the individual modules are abstracted and separated from the communication architecture. This approach greatly helps system-level designers e.g. to develop different bus systems without changing the interoperating modules. SystemC supports transaction-level modelling by using the SystemC TLM extension.

C. SystemC AMS Extension

Analog and continuous signals are everywhere on the world. SystemC AMS extension is used for modelling analogue mixed-signal behaviour in a SystemC environment. This is achieved by adding a SystemC AMS library without changing the original implementation of SystemC. With the help of this extension, system-level designers can create virtual prototypes for analog systems.

III. THE CULT FRAMEWORK

CULT is a static library with all functionalities implemented in C++. It is designed to record simulation data for different abstraction levels. Hence, it supports both, C++ and SystemC programs, including SytemC TLM and SystemC AMS extensions. Beyond the standard C++ data types such as *integer* or *double* values, the framework also supports logging and tracing of specific SystemC data types, e.g. *sc_signal<>*.

A. Overview

Figure 1 gives an overview of the CULT library architecture. The base functions of the library form the level 1 (LV1), which supports the logging and tracing of standard C++ data types. Level 2 (LV2) can be either the SystemC extension and/or user defined extensions. The SystemC extension extends the support of the framework to SystemC signals and ports. Level 3 (LV3) can either base on the SystemC extension or user defined extensions. Additional extensions based on the LV2 SystemC design can extend the support of CULT to SystemC TLM sockets and/or SystemC AMS ports. CULT also supports user defined extensions based on LV3, which makes the framework highly extensible to further standards or use cases. The configuration of the library is achieved by using external XML files.

As the current OSCI SystemC kernel does not support the execution of multiple OS-threads, it does not benefit from multi-core architectures. To increase the performance of the simulation with CULT, the logged and traced simulation data are written out periodically by the framework using POSIX threads. The frequency of the outsourcing can be configured

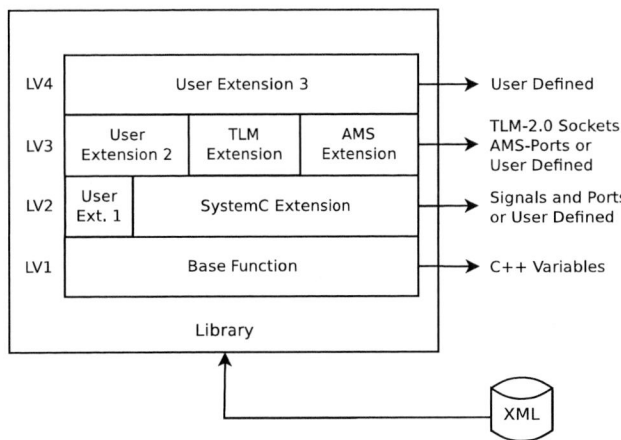

Fig. 1: CULT Library Overview

by the user, so that it can be adapted to specific use cases to achieve the best performance.

One important property of CULT is the "time-shift" feature, which allows the user to backtrack simulation data. To achieve this, CULT maintains a simulation data buffer. The buffer stores the simulation data for a given time interval retroactively to the simulation time. Once needed, e.g. in case of a special event, the user can activate this buffer and write out the content in it. An important use case of this feature is the interconnection with assertion-based verification (ABV). The user can employ this feature to write out simulation data before the assurance of the assertion failed. This gives more information for subsequent cause-seeking.

B. Configuration

CULT allows a flexible and dynamic configuration of the logging and tracing behaviour by external XML configuration files with a predefined structure as shown in Listing 1.

Listing 1: CULT Configuration File

```
1  <config
      xmlns:xsi="http://www.w3.org/2001/XMLSchema-instance">
2    <log_switch>1</log_switch>
3    <log_mode>cpp</log_mode>
4    <log_level>
5      <level>CULT_DEBUG</level>
6      <level>CULT_INFO</level>
7      <level>CULT_TRACE</level>
8      ...
9    </log_level>
10   <use_parameter>default</use_parameter>
11   <log_output_interval unit="SEC">10</log_output_interval>
12   <default>
13     ...
14   </default>
15   <module_list>
16     <module>
17       ...
18     </module>
19     ...
20   </module_list>
21 </config>
```

CULT uses the open source C++ parser TinyXML [6] to parse configuration files. If the user creates such a configuration file and tells CULT the explicit path of it, all parameters in it will be loaded. Otherwise, default values will be used. This mechanism allows the user to dynamically change the logging and tracing parameters at runtime by loading a different configuration file. The parameters that can be defined are:

- **log_switch**: enables or disables the logging and tracing library.
- **log_mode**: determines the log mode of the framework. CULT has two log modes: C++ mode and SystemC mode. C++ mode is considered for normal C++ programs and SystemC mode for SystemC programs (including programs using the TLM and/or AMS extensions). The difference between this two modes is the time format. C++ mode uses operating system time and SystemC mode the SystemC simulation time.
- **log_level**: each object that is going to be logged has a log-level. This entry defines which log-levels are activated for logging and tracing. The library predefines six log-levels: *CULT_DEBUG, CULT_INFO, CULT_WARNING, CULT_ERROR, CULT_FATAL* and *CULT_TRACE*.
- **use_parameter**: the framework considers this entry only in "systemc" log mode. Value "default" tells the framework to use parameters in the <default> block of the configuration file. All modules of the SystemC program will then use the same parameters in this block. Value "module_list" will cause the framework using the parameters in the <module_list> block of the configuration file. This allows the user to configure each SystemC module with different parameters. Note that modules that are not listed in the <module_list> block will use the parameters in the <default> block. Hence the <default> block is not optional, it should always be given.
- **log_output_interval**: this entry is optional. Complex systems have a huge amount of simulation data to be recorded. In order to avoid data loss (unexpected simulation interruption) or a memory overflow, the recorded data should be written out regularly. The user can specify the interval here. If this entry is not specified, the framework will use 1 second OS time in "cpp" mode and 1 second (SC_SEC) SystemC simulation time in "systemc" mode as default.
- **default**: this is the <default> block mentioned above. Listing 2 shows the structure of this block. The parameters that can be defined in this block are:

Listing 2: <default> Block

```
1   <default>
2     <log_file_mode>csv,vcd</log_file_mode>
3     <console>on</console>
4     <log_level>CULT_TRACE</log_level>
5     <log_begin_time unit="SEC">0</log_begin_time>
6     <log_duration unit="SEC">20</log_duration>
7     <log_max_value>100000</log_max_value>
8     <log_min_value>1</log_min_value>
9     <log_backwards unit="SEC">
        30</log_backwards_passive>
10  </default>
```

- *log_file_mode*: this parameter determines the file type to store log data. CULT supports *txt*, *csv* and *vcd* files to record the logging and tracing data. Listing 3 shows an example of a csv log file.

Listing 3: File Output (csv)

```
int_example,9,none,CULT_DEBUG,2012-03-07 17:04:07
int_example,50,none,CULT_DEBUG,2012-03-07 17:04:08
...
port_example1,R,5,Module_1,CULT_TRACE,50 ns
port_example2,W,10,Module_2,CULT_TRACE,100 ns
port_example2,W,15,Module_2,CULT_TRACE,200 ns
...
socket_example,initiator,CULT_TRACE,BLOCKING,
    (none),10 ns,TLM_WRITE_COMMAND,
    28,4,40,TLM_OK_RESPONSE,20 ns
socket_example,initiator,CULT_TRACE,BLOCKING,
    (none),10 ns,TLM_WRITE_COMMAND,
    2c,4,44,TLM_OK_RESPONSE,30 ns
...
```

- *console*: CULT supports the output of logging and tracing data to console. This parameter turns this feature on or off. Listing 4 shows an example console output of a SystemC TLM-2.0 log socket.

Listing 4: Console Output

```
<--------INITIALIZE TLM LOG SOCKET-------->
SystemC Time: 0 s
Variable Name: socket_example
LOG-Level: CULT_TRACE
Module Name: initiator
<----------------------------------------->
...
<---------------LOG ENTRY BEGIN----------->
SystemC Time: 0 s
TLM Socket Name: socket_example
Transport Type: BLOCKING
Delay: 10 ns
****** Transaction ******
TLM_WRITE_COMMAND
Address: 20
Number of Bytes: 4
Payload: 32
TLM_OK_RESPONSE
************************
LOG-Level: CULT_TRACE
Module Name: initiator
<----------------LOG ENTRY END------------>
...
```

- *log_level*: optional parameter. Determines the log-level. The log-level can be changed using macros at simulation runtime.
- *log_begin_time*: optional parameter. This parameter determines when the framework begins to record simulation data.
- *log_duration*: only considered if log_begin_time is defined. Combined with the begin time, the logging and tracing interval can be determined.
- *log_max_value*: optional parameter. Filter function of the framework. Determines the upper threshold value of the recorded simulation data.
- *log_min_value*: optional parameter. Determines the lower threshold value of the recorded simulation data.
- *log_backwards*: optional parameter. Enables "time-shift" function of CULT. It defines a time value. CULT will then manage a buffer of simulation data for this given time period retroactively to the present time. With this feature enabled, CULT allows the user to backtrack simulation data for this given time, e.g. by occurrence of a special SystemC event.
- **module_list**: this block in the configuration file allows the

62

user to configure each SystemC module separately. The structure is shown in Listing 5. The parameter structure of each module is exactly the same as an <default> block. The only difference is that the module name must be given (line 4). Simulation modules not existing in this list will use the content in the <default> block.

Listing 5: <module_list> Block

```
1  <module_list>
2    <module>
3      <name>INSERT_MODULE_NAME_HERE</name>
4      <log_file_mode>none</log_file_mode>
5      <console>on</console>
6      <log_level>CULT_TRACE</log_level>
7      <log_begin_time unit="SEC">0</log_begin_time>
8      <log_duration unit="SEC">20</log_duration>
9      <log_max_value>1000</log_max_value>
10     <log_min_value>-30</log_min_value>
11     <log_backwards unit="SEC">10</log_backwards>
12   </module>
13   ...
14 </module_list>
```

C. C/C++ Logging and Tracing

CULT supports logging and tracing functionalities for normal C++ programs without a SystemC environment. Listing 6 shows a short example how CULT is applied in this case.

Listing 6: CULT C++ Example

```
1  #define LOGGER_ON
2  #include "log.h"
3  ...
4  int main(int argc, char* argv[])
5  {
6    CULT_VAR(int) int_example;
7    CULT_VAR_INIT(int_example, "example", CULT_DEBUG);
8
9    int_example = 50; //implicit tracing/logging
10   CULT_LOG(int_example); // explicit tracing/logging
11   std::cout << int_example << std::endl;
12
13   CULT_TEXT(CULT_DEBUG, "DEBUG_text.");
14   CULT_BACKWARD;
15   CULT_STOP;
16   CULT_START;
17   CULT_CONFIG("path/to/config/file/config.xml");
18   ...
19 }
```

The #define in line 1 is the global switch for the logging and tracing framework. CULT can be turned on by just using this single #define. If the user removes this line, not a single code in the CULT framework will be executed, so that no software overhead is added to the original source code of the program. Line 6 shows the macro used to declare C++ variables that need to be logged or traced. The type of the variable is given inside the small brackets of the macro. This is using the unitized<> [7] technique. In this case, a unitized<> object of type *int* is declared. If CULT is active, this object will keep track of the value depending on the configuration and writes out the value periodically if necessary. If the #define is not set, the object will behave just as a normal C++ integer. Line 7 then is the initialization macro of the logging and tracing integer. It sets value, name and log level of the integer. This will determine how CULT handles the values of this integer. In line 9, a normal value assignment code is shown. With CULT activated, the framework will handle the tracing, logging and outsourcing of the value automatically in the background. Line 10 then is an explicit logging macro. This macro tells CULT to record the value of the integer at this moment, even if it does not satisfy the defined logging and tracing configuration. In line 11, the integer is written to console using the stream operator.

From line 13 to line 17, a number of macros are executed. Line 13 shows the ability of CULT to log and trace freetext. The text is given together with the log level, so that only the text with the configured levels are recorded. Line 14 shows the macro to activate the "time-shift" feature presented in the configuration description. Line 15 and 16 allow the user to deactivate and reactivate CULT for a specific piece of code. This allows the user to turn off and on CULT during runtime. Finally, line 17 shows the macro, where the user can dynamically change the complete configuration of CULT at runtime. Note that all macros can only achieve their functionalities with the #define in line 1 set.

Normally, the user only needs to change four lines of the original code to integrate CULT. First, the #define should be added as shown in line 1. Second, the #include "log.h" is required. Finally, the variable definition line in the original code should be substituted with the definition and initialization macro of CULT. This allows minimal-invasive intervention of the users original code.

D. Standard SystemC Logging and Tracing

The usage of the logging and tracing library in a SystemC environment is similar to the pure C++ case. The functionality of the framework is also achieved using macros. Listing 7 shows an example code.

In line 6 and 7, a sc_in<> and a sc_signal<> with logging and tracing support are defined. The type of the data is defined inside the brackets of the macro. Before using the port and the signal, they should be initialized with their names, log-levels and the module names where they are defined. This is done in the constructor from line 16 to 20. After the initialization, the ports or signals can just be used as standard SystemC ports or signals. The values are automatically recorded in the background depending on the framework configuration. To record a value, even if it doesn't satisfy the defined logging and tracing configuration, the explicit logging macro SC_LOGGER_LOG can be used (line 12).

Listing 7: CULT SystemC Example

```
1  #define LOGGER_ON
2  #include "log.h"
3  #include <systemc.h>
4  ...
5  SC_MODULE(module){
6    CULT_SC_IN(bool) port_example;
7    CULT_SC_SIGNAL(int) signal_example;
8    ...
9    void thread(){
10     port_example->default_event();
11     std::cout << port_example.read() << std::endl;
12     CULT_LOG(signal_example);
13     signal_example = 80;
14     ...
```

```
15    }
16    SC_CTOR( module ){
17        CULT_PORT_INIT( port_example , "testport",
                          CULT_TRACE, name());
18        CULT_SC_SIGNAL_INIT( signal_example ,
                          "testsignal",
                          CULT_DEBUG, name());
19        SC_THREAD( thread );
20    }
21  };
```

The framework supports the following SystemC specific data types:

- sc_signal<T>, sc_buffer<T>
- sc_port<sc_signal_in_if<T>, 1>
- sc_in<T>, sc_out<T>, sc_inout<T>

By using the corresponding macros for these data types, the framework can automatically log or trace the data in the background. Note that the framework doesn't support the logging and tracing for the general sc_port<> type. This is technically unachievable, because each interface associated with the port has its own methods. The framework hence has no knowledge of user defined methods or new interfaces with new methods in the future. But the framework can be extended for each interface type, allowing users to add logging and tracing support to their own interfaces associated with sc_port<>.

E. SystemC TLM Logging and Tracing

The framework supports the logging and tracing of SystemC TLM-2.0 sockets. The framework will log and trace the sent or received TLM-2.0 transactions of the specified socket. Listing 8 shows an example of code.

Listing 8: SystemC TLM Example

```
1   #define LOGGER_ON
2   #include "log.h"
3   #include "systemc.h"
4   ...
5   SC_MODULE( Initiator )
6   {
7     CULT_TLM_SIMPLE_INITIATOR_SOCKET( Initiator )
                                        socket_example ;
8     SC_CTOR( Initiator ): socket_example("testsocket")
9     {
10      CULT_TLM_SOCKET_INIT( socket_example , "testsocket",
                          CULT_TRACE, name()); //alternative 1
11      CULT_TLM_SOCKET_INIT_WITH_CALLBACK( socket_example ,
                          "testsocket", CULT_TRACE,
                          name(),&int_payload ); //alternative 2
12      CULT_TLM_SET_DBG( socket_example , 1);
13      SC_THREAD( thread_process );
14    }
15    void thread_process ()
16    {
17      tlm::tlm_generic_payload* trans =
                          new tlm::tlm_generic_payload ;
18      ...
19      SC_STREAM(* trans );
20      socket_example->b_transport( *trans , delay );
21      ...
22    }
23    ...
24  };
```

The principles of the logging and tracing are the same as the logging of SystemC ports. The TLM-2.0 sockets should be first defined and then initialized in the constructor of the SystemC module. Line 7 shows the definition and line 10 to 12 the initialization. Note that line 10 and 11 are

different initialization macros. Only one of them is needed. The difference of these two initialization methods is the support of callback methods. The macro in line 10 doesn't need to specify a callback method. By using this macro, the framework will output the generic payload of all TLM-2.0 transactions formatted as hex. In line 11, the macro needs the user to provide a reference to a callback method to handle the data stored in the generic payload. Using this callback method, the user can output the payload in the right data format.

Line 12 is a configuration macro. The framework doesn't record TLM-2.0 transactions sent with the debug transport method *transport_dbg()* by default. In most times, these transactions don't need to be logged or traced. But if required, the user can turn on logging and tracing for this method by using this macro with the second parameter set to 1. If the second parameter is 0, the macro will turn off the logging and tracing of those transactions.

The logging and tracing framework supports the output of TLM-2.0 transactions with stream operators to console. The macro for this purpose is given in line 19 of Listing 8.

F. SystemC AMS Logging and Tracing

CULT supports the logging and tracing of SystemC AMS *sca_in<>* and *sca_out<>* ports. A simple example of the usage of the framework for *sca_out<>* is shown in listing 9. For SystemC AMS, the framework follows the same principle. The logging and tracing port is first defined and then initialized. The definition is given in line 7 and the initialization in line 15. Recording the simulation data is then done automatically in the background with the user defined configuration. The framework can be extended by the user to add logging and tracing functionalities to custom user SystemC AMS ports.

Listing 9: CULT SystemC AMS Example

```
1   #define FZI_LOGGER_ON
2   #include "log.h"
3   #include <systemc-ams.h>
4   ...
5   SCA_TDF_MODULE( signal_producer )
6   {
7     CULT_TDF_OUT( double ) ams_example ;
8     ...
9     void set_attributes ()
10    {
11      ams_example.set_timestep( sc_get_time_resolution() );
12      ...
12    }
13    void initialize ()
14    {
15      CULT_AMS_PORT_INIT( ams_example , "ams_testport_1".
                          CULT_TRACE, name());
16      ...
17    }
18    void processing ()
19    {
20      ams_example.write( sc_time_stamp().to_seconds());
21      ...
22    }
23    SCA_CTOR( signal_producer ){}
24  };
```

IV. APPLICATION AND PERFORMANCE

A. Assertion-Based Verification(ABV) Interconnection

CULT can be extended by ABV [5] to support design testing and validation in early design phases. Listing 10 shows a simple example of ABV. The code defines two integers and two propositions. Depending on the condition, an accept or reject callback function is executed.

Listing 10: ABV Example

```
1   void TestModule::thread()
2   {
3     int a = 0. b = 1;
4     //declare reference for local variables
5     ABV_DECLARE_REFERENCE(A, a);
6     ABV_DECLARE_REFERENCE(B, b);
7     //declare two contradicting propositions
8     ABV_DECLARE_PROPOSITION("a_ne_b", A!=B);
9     ABV_DECLARE_PROPOSITION("a_eq_b", A==B);
10    //declare properties to test callback for
11    //accepted//rejected properties
12    sc_monitor_object *accept =
          ABV_TRIGGERED_MONITOR("F_a_ne_b", trigger_event);
13    sc_monitor_object *reject =
          ABV_TRIGGERED_MONITOR("G_a_eq_b", trigger_event);
14    //function style callbacks
15    reject->add_reject_callback(reject_callback_function);
16    accept->add_accept_callback(accept_callback_function);
17    ...
```

The interconnection of CULT with ABV is achieved by adding CULT macros inside the accept or reject callback functions. Listing 11 shows an interconnection of CULT with ABV in the reject callback function. The user can then activate CULT on reject condition (line 4) or activate the "time-shift" function (line 5). This greatly supports the user in testing and validation.

Listing 11: Interconnection with ABV Example

```
1   static void
       reject_callback_function (sc_monitor_object *monitor)
2   {
3     ...
4     CULT_START;
5     CULT_BACKWARD;
6     ...
7   }
```

B. Performance Impact

The software overhead of CULT is generated by maintaining and outsourcing simulation data. Two applications are analysed in order to show the performance impact.

The first application is a JPEG encoder implemented in SystemC. In the simulation scenario, it takes a 2.1MB pgm file as source for the encoding. The second application is a traffic sign recognition system implemented in SystemC. It takes a video as a source file and recognizes the traffic signs in it. The simulation time of the second application is limited to 80 seconds of SystemC time.

The two applications are run 50 times each with and without CULT. The average runtime needed for the simulation, is then calculated from the measured real time of each run. The results are shown in Table I. It can clearly be seen that the runtime overhead of CULT depends on the amount of the recorded simulation data. But even with 2 million recorded data in the JPEG application, CULT produces only 0.008% time overhead.

V. CONCLUSION AND FUTURE WORK

In this paper, we introduced CULT which is a novel and powerful framework for tracing and logging C-based designs of embedded hardware/software systems. It is an efficient

Application	JPEG Encoder	Traffic Sign Recognition
Number of Log Entries	~ 2 million	~ 25000
Size of Log Data	73 MB	1.1 MB
Runtime with CULT	139.231 sec	34.530 sec
Runtime w/o CULT	138.080 sec	34.263 sec

TABLE I: CULT Performance Impact

framework to be used at different abstraction levels. Both, C++ and SystemC applications, including the SystemC TLM and SystemC AMS extensions are supported by CULT. It is flexible, configurable, and provides several powerful features. We believe that CULT satisfies the need of a unified and powerful logging and tracing framework for C-based designs. However, the work described in this paper is just the first milestone. Additional features like the integration of the framework configuration by IP-XACT [8] system designs are under examination. This paper documents the architecture and first steps of the project. We distribute the code as an open source tool under the Apache 2 license. The download link is [9]. Please note that the implementation of the described work in this paper may change before its first release due to additional or modified functionalities.

ACKNOWLEDGMENT

This work has been developed in the project SANITAS. SANITAS (project label: 01 M 3088) is partly funded by the German ministry of education and research (BMBF) within the Research Program ICT 2020.

REFERENCES

[1] *IEEE Standard SystemC Language Reference Manual.* IEEE, 9 2012.

[2] L. Cai and D. Gajski, "Transaction Level Modeling: An Overview," in *Proceedings of the 1st IEEE/ACM/IFIP international conference on Hardware/software codesign and system synthesis*, ser. CODES+ISSS '03. New York, NY, USA: ACM, 2003, pp. 19–24. [Online]. Available: http://doi.acm.org/10.1145/944645.944651

[3] K. Einwich, "Introduction to the SystemC AMS extension standard," in *Design and Diagnostics of Electronic Circuits Systems (DDECS), 2011 IEEE 14th International Symposium on*, April 2011, pp. 6 –8.

[4] "SystemC Tracing Library by LogicPoet." [Online]. Available: http://www.logicpoet.com/tracer

[5] R. Weiss, J. Ruf, T. Kropf, and W. Rosenstiel, "Efficient and Customizable Integration of Temporal Properties into SystemC," in *Forum on Specification and Design Languages (FDL'05)*, Sept. 2005, pp. 271–282.

[6] "TinyXML: An Open Source C++ XML Parser." [Online]. Available: http://www.grinninglizard.com/tinyxml

[7] C. Kerstan, N. Bannow, and W. Rosenstiel, "Efficient Architecture Evaluation Using Functional Mapping," in *Languages for Embedded Systems and their Applications*, ser. Lecture Notes in Electrical Engineering, M. Radetzki, Ed. Springer Netherlands, 2009, vol. 36, pp. 167–182. [Online]. Available: http://dx.doi.org/10.1007/978-1-4020-9714-0_11

[8] "IP-XACT XML Schema." [Online]. Available: http://www.accellera.org/activities/committees/ip-xact

[9] "CULT Logging and Tracing Framework." [Online]. Available: http://www.fzi.de/sim/dienstleistungen/tools/

[10] *SystemC Version 2.0 User's Guide.* Open SystemC Initiative, 2002.

[11] "Open SystemC Initiative TLM-2.0 LANGUAGE REFERENCE MANUAL." [Online]. Available: http://www.systemc.org/downloads

[12] *SystemC-AMS*, Open SystemC Initiative, Mar. 2007.

ARMV8 DEBUG AND TRACE ARCHITECTURES

Michael Williams

ARM Limited, 110 Fulbourn Road, Cambridge, England.

ABSTRACT

The ARM® processor family is the most widely used 32-bit processor family, and has consistently included debug features, starting from the ARM7TDMI® processor [1]. Embedded debug forms part of the ARMv7 architecture [2]. The most recent addition to the ARM architecture family is ARMv8 [3], which represents the biggest change in the architecture's history. This paper looks at the impact of the new architecture on debug and trace, and reviews some of the lessons learnt from previous architectures.

Index Terms—software debugging, computer architecture.

1. INTRODUCTION

As discrete components were replaced with *Systems-on-Chip* (SoCs) during the 1990s, so the *In-Circuit Emulator* (ICE) debugger began to be replaced by debug hardware embedded into the SoC itself, often referred to as an *embedded ICE*.

ARM7TDMI is an early example of a processor with embedded ICE. The ARM7TDMI embedded ICE stops the processor clock and communicates with a host debugger on a second computer using a relatively low-cost *debug probe*. Over time the features of the embedded ICE expanded to include trace [4] and system visibility [5].

Debug probes are seldom used outside embedded markets and initial bring-up of a board or system by its manufacturer.

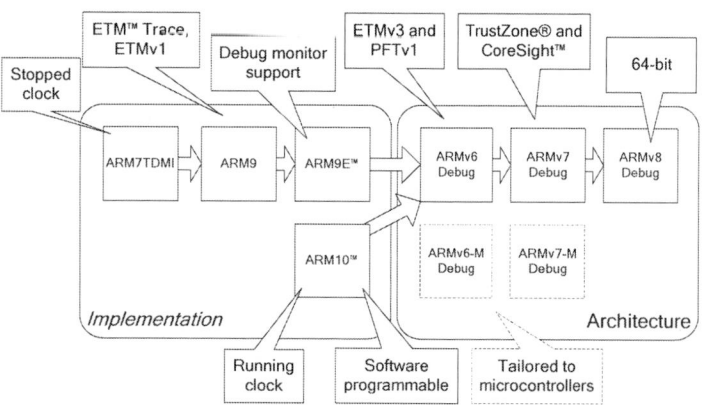

Fig. 1 Evolution of the ARM debug architecture (left to right)

Developers creating software for an *open platform* usually prefer *hosted* development; that is, where the computer which will run the software (or one very much like it) also hosts the compiler, utilities, debugger, etc.

Although first appearing in ARM10™, version 6 of the ARM architecture (ARMv6) was the first version to support *self-hosted* development: ARMv6 provides software access to the processor's debug resources and integrates debug events into the exception architecture. ARMv6 was also the first version of the ARM architecture to describe debug as an architectural feature. On earlier processors, debug was an implementation feature coupled to the micro-architecture (Fig. 1).

However, ARM processors have continued to support the embedded *external debug* model, and the ARM CoreSight™ SoC-debug platform [5] extends the software and external debug view to the entire SoC.

2. INTRODUCTION TO ARMV8

ARMv8 extends the ARM architecture by adding a 64-bit execution state (AArch64), alongside the existing 32-bit execution state (AArch32). To support 64-bit ARMv8 defines:

— A new instruction set for AArch64 (A64).
— A revised exception model for exceptions taken to AArch64 state, with fewer banked registers and modes.
— 64-bit virtual addressing in AArch64 state[1].

ARMv8 also supports Security and Virtualization features, as in the previous version of the ARM architecture, ARMv7.

For debug, AArch64 poses two main challenges:

— Integrating debug events into the new exception model.
— Supporting longer virtual addresses.

Both are relatively straightforward problems, but required re-engineering of debug and trace to fully support ARMv8. Therefore the architects took the opportunity to draw on their

---

[1] Virtual addresses in AArch64 state are actually up to 49 bits long (512TB). This is to reduce implementation costs whilst keeping space for further expansion, in around 2035.

66

experience and redefine debug and trace to provide an improved solution for the ARMv8 markets, without losing backwards compatibility for software.

Enabling self-hosted debug in secured environments

ARMv7 processors provide two *debug-modes*, supporting two types of development:
— *Monitor debug-mode*, for self-hosted debug
— *Halting debug-mode*, for external debug.

These have different security requirements:
— Self-hosted debug is an *operating system* (OS) service. The OS programs the debug hardware, which in turn generates debug exceptions processed by the OS. The OS is gate-keeper to all debug functions.
— External debug is a hardware function. The external debugger can have full visibility of the system. Therefore hardware controls are needed. These might be controlled statically by fuses or dynamically by an authentication module.

A hardware platform specification, with legitimate concerns about exposing the system to hardware attacks through debug, might require the SoC developer to provide means to block external debug using the hardware controls. However, in ARMv7 these controls also block self-hosted debug, making the SoC platform harder to use for software development on open software platforms.

ARMv8 puts self-hosted debug fully under the control of OS software. Hardware controls are provided to block external debug access, but these do not impact self-hosted debug. This is done without losing backwards compatibility for self-hosted debug software compliant with ARMv7.

Virtualization and Security features are also provided. These are largely inherited from the ARMv7 Virtualization Extensions, but include some new features to protect Secure operating software from attack by Trojan software executing in the Non-secure environment when Secure debug is enabled.

Reducing the complexity of external debug

When a debug event for an external debugger is generated, the processor halts and enters a special *Debug state*.

This special operating state behaves differently to the normal operating state of the processor, as control has passed to an external debugger. However, differences in behavior create architectural and micro-architectural complexity. A goal of ARMv8 debug is to reduce this complexity.

To allow for this, the architects recognized that, although backwards compatibility is paramount for software executing on the processor, it can be sacrificed for external debuggers. This means that companies providing debug probes for ARMv8 processors will need to develop new software. However standard debug interfaces from ARMv7 are reused.

Supporting AArch64 debug and trace

System registers containing virtual addresses, such as breakpoint and watchpoint addresses, must be extended. For self-hosted debug, the instructions that access these registers are 64-bit transfers in AArch64 state. For external debug, 64-bit registers are mapped to a pair of 32-bit locations, as the external debug interface is naturally 32-bit.

A new trace protocol has been defined for ARMv8 to support 64-bit virtual addressing, as extending existing trace protocols to support AArch64 was not possible. The architects have also addressed tracing higher-performance processors without impacting the bandwidth required for program-flow trace compared to previous protocols.

3. ARMV8 SELF-HOSTED DEBUG IN DEPTH

As stated above, ARMv8 maintains backwards compatibility in AArch32 state for ARMv7 software written to use self-hosted debug. This means ARMv8 provides all the self-hosted debug features of ARMv7 to AArch32 state:
— Hardware breakpoints and watchpoints, with linked context-ID and VMID comparators.
— Mismatch breakpoints for simple single-stepping.
— Vector catch for trapping accesses into the kernel.
— A software breakpoint instruction.
— Access to hardware debug features through system register instructions.

To reduce future complexity, use of some features is deprecated and these features are not available in AArch64 state.

Mismatch breakpoints: micro-architecture led features

Mismatch breakpoints were a novel solution to the problem of providing single-step at minimal cost. The result from a hardware breakpoint is inverted, thereby giving a breakpoint which matches every instruction in a given context, except for the instruction being stepped.

When an instruction is single-stepped in ARMv7::
— A mismatch breakpoint is programmed with the address and context of the instruction.
— Control passes from the debugger to the instruction.
— The processor executes that single instruction.

— The following instruction generates a breakpoint as it has a different address.

If the instruction generates an exception it is handled in a different context, meaning the effect is repeatable on return from the exception handler to the original instruction's context.

Mismatch breakpoints first appeared on a 32-bit processor supporting fixed-length instructions, with limited dual-issue and in-order execution. To support variable-length instructions in ARMv7, mismatch breakpoints had to be redefined. What was initially a simple solution became, over time, a complex one.

ARMv8 replaces mismatch breakpoints with hardware single-step. Although this does not re-use the existing hardware breakpoints feature, it is defined so that it captures the required *debug illusion* for single-stepping, allowing designers to use an appropriate implementation for a microarchitecture.

When an instruction is single-stepped in ARMv8:
— Stepping is enabled in a control register.
— Control passes to the instruction with a status bit in the *program status register* (PSTATE) set to 1.
— The processor executes that single instruction.
— The following instruction generates a debug event.

If the instruction generates an exception, the processor copies PSTATE to a *saved program status register* (SPSR). On return from the exception handler, the SPSR is copied back to PSTATE. This provides the illusion of *stepping over* the exception.

The ARMv8 debug architecture also supports *kernel debugging*. In this mode, the debugger can *step into* an exception handler. A debug exception mask is provided to prevent re-entrant exceptions during stepping.

4. ARMV8 EXTERNAL DEBUG IN DEPTH

ARMv8 external debug keeps many of the features from ARMv7 external debug, including the hardware breakpoints and watchpoints shared by self-hosted debug and support for debug over power-down. Additions to external debug include:
— A halt instruction.
— *Embedded Cross-Trigger* (ECT) for multi-processors.
— Hardware single-step, as for self-hosted debug.
— Reset catch and exception catch features.

Although the architects decided that it was acceptable to lose ARMv7 compatibility for external debug, reuse of existing components was an important design constraint.

Access to the ARMv8 processor for external debug is made through a 32-bit CoreSight *Debug Access Port* (DAP) interface, as used by ARMv7 processors. This hardware reuse in turn allows reuse of the low-level software drivers for IEEE 1149.1 "JTAG" and *Serial-Wire Debug* (SWD) debug ports. The access sequences for these interfaces are unchanged for ARMv8.

Debug state: the cost of complexity

In the ARM architecture, when a debug event is generated for an external debugger, the processor halts and enters a special *Debug state*. In this state, instructions are not fetched from memory, but instead the external debugger writes instructions to an *Instruction Transfer Register* (ITR), which the processor executes.

In ARMv7, any valid instruction can be executed in Debug state. This gives a simple mechanism for the debugger to access any state of the processor. However, instructions cannot normally access all processor state all of the time. For example:
— The architecture supports a hierarchy of privilege levels and blocks visibility at some privilege levels.
— Some processor state is never directly visible. For example, some *Current Program Status Register* (CPSR) fields are only visible by the effect they have, and in an SPSR after taking an exceptions.
— The *program counter* (PC) is only valid for instructions fetched from memory.

ARMv7 defines altered behavior for instructions that access this state in Debug state. This adds complexity, and requires that many parts of the processor are aware of the current state.

ARMv8 reduces this complexity by adding specific instructions to access processor state that is not normally visible. In addition, ARMv8 restricts which other instructions can be executed in Debug state, to reduce the overall validation space.

External debug is a key feature for the bring-up of new systems, meaning it is important that it is right first time.

Halt instruction

The ARMv7 software breakpoint instruction is modal: depending on the configured *debug-mode* of the processor, it generates either a debug exception (self-hosted) or entry to debug state.

This modal behavior has led some to be wary of its use. An instruction that might halt the processor is not necessarily a desirable feature in an OS environment. Although for the

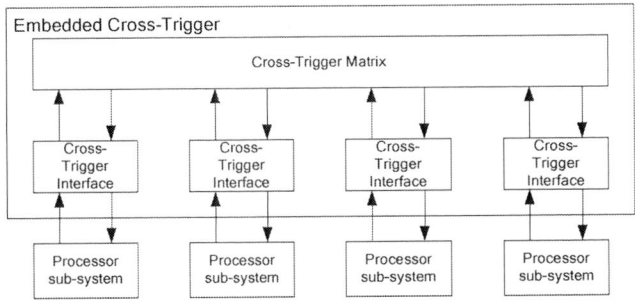

Fig. 2: Embedded Cross-Trigger

ETMv4 trace element	A	R	M
Conditional branches (program-flow)	●	●	●
Exceptions	●	●	●
Variable cycle counting	○	○	○
Variable global time-stamping	○	○	○
Event tracing	○	○	○
Conditional *non-branch* instructions	-	○	○
Loaded data trace	-	○	○
Stored data trace	-	○	○

Table 1: ETMv4 feature set (●=always on, ○=programmable)

instruction to halt the processor, the processor must first be programmed into Halting debug-mode by an external debugger, it was straightforward to add a separate halt instruction in ARMv8. The existing software breakpoint instruction can only generate a debug exception.

If halting is either not enabled or not allowed by the security model, the halt instruction generates an Undefined Instruction exception.

Embedded Cross-Trigger: architectural consolidation

An *Embedded Cross-Trigger* (ECT, Fig. 2) comprises one or more *Cross-Trigger Interfaces* (CTIs) that interface to processors (or other components) and map processor events to ECT events, and a *Cross-Trigger Matrix* (CTM) that transports ECT events between CTIs.

For example, the ECT can be configured to halt all processors in the system when any one processor halts due to a debug event. The ECT can also be used events to inject debug under debugger control.

To reduce variability, given that multi-processor systems are prevalent for ARMv8, the architects decided to require that a CTI is always present.

5. ETMV4 IN DEPTH

There are four previous generations of program-flow trace for the ARM architecture:

ETMv1 (*Embedded Trace Macrocell, version 1*), for ARM7TDMI and ARM9™-family processors, trace the *pipeline status* on each *cycle*. This generates large amounts of trace and supports only simple pipelines.

ETMv2 extends ETMv1 for ARM10-family processors. The ARM10 has a longer pipeline with additional pipeline status codes.

ETMv3 traces each *instruction* (as opposed to each *cycle*) and data access retired by the processor, in order. The trace is compressed and output as a byte packet protocol to CoreSight.

PFTv1 (*Program-flow Trace, version 1*) is similar, but only *waypoint* instructions (branches, etc.) are traced. Data is not traced. Further compression techniques improve trace efficiency.

Extending trace ETMv3 or PFTv1 to support 64-bit addresses was not possible. In addition, ETMv3 and PFTv1:
— Assume no speculation of instructions.
— Support only shallow out-of-order data transfers (ETM only).
— Have a rich programmers' model with many optional features.

Some implementations of ARMv7-R also that require data trace but are not suitable to use ETMv3. For example Cortex-R7 [6] is a high-performance processor for real-time applications that implements out-of-order execution.

Rather than define two new trace architectures, the architects realized it was possible to define a single new trace architecture with different profiles for distinct target markets.

Thus **ETMv4** has been defined to support:
— High performance, including speculative execution.
— Data trace, including out-of-order data (ARMv7, real-time and microcontroller profiles only).
— 64-bit virtual addressing (ARMv8 only).

Trace profiles

The ARM architecture defines three processor profiles:
— Application (A) processors are typically high-performance multi-processors. Basic program-flow trace is of interest. Data trace is interesting, but not feasible at these performance points. *Note:* ARMv8 supports only the A-profile.
— Real-time (R) and microcontroller (M) processors are typically at a lower performance point (though not *low performance*). Data trace is an absolute requirement for some of these processors.

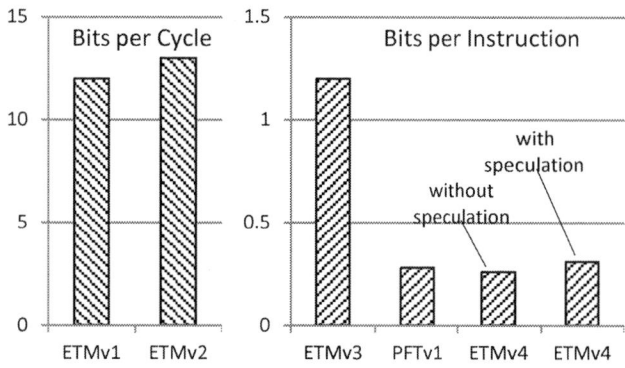

Fig. 3: Trace protocol bandwidth comparison

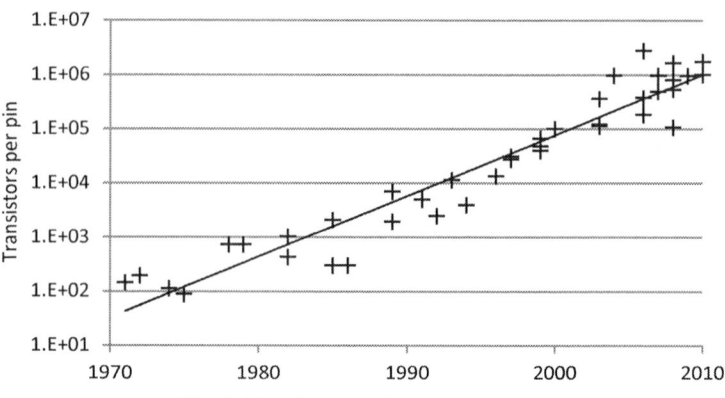

Fig. 4: Changing rate of transistors per pin

To address these different markets, ETMv4 supports a range of features but two profiles: one for the A-profile, and one for the R and M profiles. The set of supported trace elements varies by profile (Table 1).

Basic program-flow trace in ETMv4

Program-flow trace in ETMv4 is based on tracing only *waypoint* instructions, as in PFTv1. If data is not being traced, then branch instructions are the only waypoint instructions.

Branch prediction in modern processors is highly accurate, but occasionally causes speculative execution along the wrong program path at a waypoint, leading to the need to unwind execution and take the correct path.

In ETMv4 waypoints can be traced speculatively, whereas in PFTv1 only committed waypoints are traced. This means the ETM does not have to buffer all the information about a waypoint to its commitment point. *Commit* and *cancel* elements are output in the trace stream to resolve speculation.

The cost of removing speculative paths is pushed onto the debugger decoding the trace stream, and results in a small decrease in efficiency (Fig. 3). But the benefit comes in reduced area and power generating the trace.

The trace protocols are defined such that speculation can be resolved without deep inspection of the trace. In particular, unlike full trace decode, speculation can be resolved without reference to the program image.

Data trace in ETMv4 *(R and M profiles only)*

Data trace is linked to waypoint trace using timestamps and *key-pairs*. Full instruction and data trace consists of three linked elements:

P0 Waypoints. For data trace, load and store instructions are waypoints.

P1 Data addresses. These can be traced out of order and are linked to a P0 element by a key. If a P0 element is canceled, all its P1 elements are canceled.

P2 Data values. These can also be traced out of order and are linked to a P1 element by a key. If a P1 element is canceled, all its P2 elements are canceled.

This allows the instruction and data trace streams to be handled independently of each other in hardware, providing more opportunities for compression. P0 elements appear in the instruction trace stream, P1 and P2 elements in the data trace stream.

Protocol efficiency

It is important that each trace protocol generation maintains or improves efficiency as CPU frequency and core count are multiplicative factors that outstrip increases in external trace port bandwidth. Fig. 3 illustrates trace bandwidth numbers for different trace protocol versions. The following points are of note:

— By moving from *per-cycle* to *per-instruction* trace, ETMv3 improved efficiency by an order-of-magnitude
— PFTv1 gave a further improvement by tracing waypoints instead of every instruction and using further compression techniques
— ETMv4 maintains this efficiency, even when tracing speculatively executed instructions.

Note: These values are indicative only. Trace bandwidth requirements will vary according to the software being traced, the tracing features enabled, and the microarchitecture of the processor.

Such improvements in trace efficiency come at a cost of increased complexity on-chip. However, as the number of pins on packages has not kept pace with Moore's Law for the number of transistors, this cost can be justified (Fig. 4). The number of pins available for trace actually decreases as in-

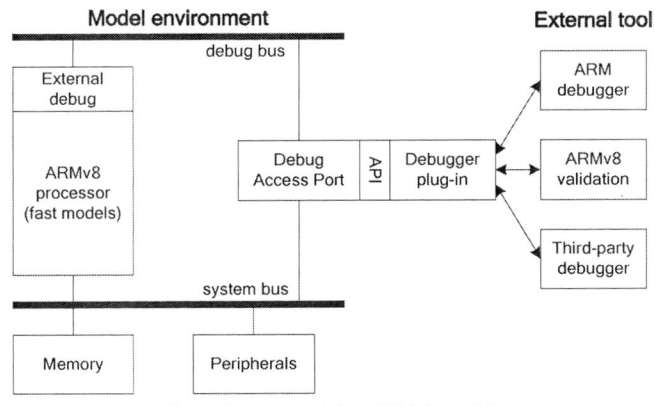

Fig. 5: Structure of the ARMv8 model

creased transistor density pushes more functions onto a single chip.

6. MODEL-LED DEVELOPMENT

ARM has created a software model of an ARMv8 processor, based on ARM *fast models* technology. In addition to an ARMv8 processor model and the usual system components of a *virtual platform* (VP), the model has been extended to include:

— Models of the ARMv8 processor's external debug.
— A plug-in API for CoreSight-like access to the system from an external debugger.

Fig. 5 shows the structure of the model.

The API allows development of a variety of different plug-ins. Plug-ins are dynamically loaded and configured by the models platform. Plug-ins developed to date include:

— Debugger plug-ins, to allow an external debugger to connect to the model as if it were real hardware.
— Validation plug-ins, used for architectural validation.

This structure allows debug tools to be developed ahead of silicon. A tool can be developed and tested against a reference model known to be architecturally compliant.

The debug API is not included in released versions of the model. Tools developers can contact ARM Limited for more information.

7. CONCLUSIONS

ARMv8 represents the biggest change to the ARM architecture in its history.

The ARM architects took this opportunity to selectively redefine aspects of the debug architecture, maintaining back-

wards compatibility for ARMv7 processors whilst extending the capabilities for the future.

The focus of debug has shifted from *external debug* support to *self-hosted debug* support. This provides better support for general-purpose computing platforms. External debug provides effective support for bringing up systems, whereas self-hosted debug allows software developers to debug operating systems and applications at low cost.

There is a future-legacy cost associated with architecture, meaning it is important that architectures provide crisp semantics for features rather than describing implementations. Good ideas in one generation can become a burden on future implementations.

ETMv4 extends the roadmap of ETM trace protocols by providing support for speculative execution, blending the high efficiency of PFTv1 instruction trace with support for out-of-order data trace, as needed for embedded systems.

Model-led development has allowed the architects to design these features in conjunction with developing tools ahead of silicon, reducing the time to market.

8. REFERENCES

[1] ARM Ltd, *ARM7TDMI Technical Reference Manual*. Cambridge, England, 1994-2004.

[2] ARM Ltd, *ARM Architecture Reference Manual ARMv7-A and ARMv7-R edition*, Issue C.b ed. Cambridge, England, 2012.

[3] R. Grisenthwaite. (2011) ARM Ltd. [Online]. http://www.arm.com/files/downloads/ARMv8_Architecture.pdf

[4] ARM Ltd, *Embedded Trace Macrocell Architecture Specification*. Cambridge, England, 1999-2011.

[5] ARM Ltd, *CoreSight Architecture Specification Rev 1.0*. ARM IHI 0029B, 2005.

[6] ARM Limited. (2012) Cortex-R7 Processor. [Online]. http://www.arm.com/products/processors/cortex-r/cortex-r7.php

9. PROPRIETARY NOTICE

ARM is a registered trademark of ARM Limited. The ARM logo, AMBA, Cortex and CoreSight are trademarks of ARM Limited. All other products or services mentioned herein may be trademarks of their respective owners.

9781467324540